Scientific Theories and Philosophical Stances

Lauener Library of Analytical Philosophy

Edited by
Alex Burri and Michael Frauchiger

Volume 6

Scientific Theories and Philosophical Stances

Themes from van Fraassen

Edited by
Claus Beisbart and Michael Frauchiger

DE GRUYTER

ISBN 978-3-11-101963-5
e-ISBN (PDF) 978-3-11-101980-2
e-ISBN (EPUB) 978-3-11-102025-9
ISSN 2198-2155

Library of Congress Control Number: 2024931165

Bibliographic information published by the Deutsche Nationalbibliothek
The Deutsche Nationalbibliothek lists this publication in the Deutsche Nationalbibliografie; detailed bibliographic data are available on the Internet at http://dnb.dnb.de.

Printing and binding: CPI books GmbH, Leck

www.degruyter.com

Brief Book Overview

This volume collects original research articles that engage with themes from Bas van Fraassen's seminal contributions to philosophy of science. The articles originate from talks that were presented at the 8th International Lauener Symposium on Analytical Philosophy honouring van Fraassen, the 2018 awardee of the Lauener Prize for an Outstanding Oeuvre.

Following an introduction by Matthias Egg and a laudation by Steven French, the bulk of the book comes in three parts. Part I is focused on scientific representation and opens with an extended essay that van Fraassen himself has contributed, titled "The Semantic Approach, After 50 Years". This essay offers a unique opportunity to learn about the semantic view of theories from a philosopher who has been pivotal for the development of this view. In his "Representation and Realism: On Being a Structuralist All the Way (Up and) Down", Steven French argues that his structuralist realism can meet a challenge that van Fraassen has set for accounts of scientific representation. Taking inspiration from van Fraassen, Michela Massimi, in her "The Perspectival Nature of Scientific Representation", makes a new suggestion on how the perspectivity of representations can be understood.

Part II is devoted to van Fraassen's constructive empiricism, a position that is an important opponent to scientific realism. The first chapter in this part is written by Finnur Dellsén, the 2018 winner of the Lauener Prize for Up-and-Coming Philosophers. In his "Expanding the Empirical Realm: Constructive Empiricism and Augmented Observation", he develops a new objection against constructive empiricism. In "Aesthetic Virtues and Theory Acceptance", Milena Ivanova supports van Fraassen in his doubts on the view that appeal to aesthetic virtues can cut some ice for scientific realists. In the final chapter of this part, "Realism and Instrumentalism: The Best of Both", Michael Friedman uses the example of Newton's *Principia* to argue that scientific realism and instrumentalism should be combined.

Part III of the book moves on to the question of what kind of attitude we should adopt towards science and towards metaphysics. Matthias Egg's chapter "Stances and Doctrines in Scientific Metaphysics" is based upon van Fraassen's proposal to conceive of philosophical views as stances rather than doctrines. The chapter examines how this proposal has fared in recent work on scientific metaphysics. Finally, Nancy Cartwright, in her "Evidence, Relevance and Warrant: In Defence of Voluntarism", provides a new defence of van Fraassen's voluntarism.

The volume concludes with Bas van Fraassen's "Reply to the Conference Participants".

https://doi.org/10.1515/9783111019802-001

Acknowledgements

Our first thanks go to Bas van Fraassen, the winner of the Lauener Prize for an Outstanding Oeuvre in Analytical Philosophy 2018. His invaluable contributions to the symposium, his friendly and constructive engagement with the commentators and his help behind the scenes in staging the symposium are appreciated with great thanks. We are particularly glad that he has contributed an extensive discussion of the Semantic View of Theories, roughly 50 years after it came to life. We hope that this paper will become a landmark reference for philosophical work on theories. We are also grateful for the excellent cooperation with him during the edition of this volume.

We'd also like to thank our contributors for their commentaries on Bas van Fraassen's work and for elaborating their ideas further in the chapters of this book. We enjoyed the collaboration with them very much. Special thanks go to Matthias Egg who has not only written a chapter for this volume but also contributed an introduction.

We extend our thanks to our colleagues from the Lauener Foundation Board, its former President Dagfinn Føllesdal and its present President Alex Burri as well as Dieter Jordi and Markus Stepanians. Their continuous support of the project has been very helpful indeed. We'd also like to thank Alex Burri in his capacity as the series co-editor.

This volume would not have been possible without Layla Huber's proof-reading and editing, for which we are extremely grateful. We'd also like to thank our partners at De Gruyter, in particular Christoph Schirmer and Anne Hiller, as well as the copy editor Konrad Vorderobermeier and Matthias Wand for the excellent collaboration.

Michael Frauchiger and Claus Beisbart, Berne, January 2024

https://doi.org/10.1515/9783111019802-002

Acknowledgements

Table of Contents

Abbreviations

AA	Akademieausgabe (*Kant's gesammelte Schriften*)
BINP	Bangladesh Integrated Nutrition Programme
DDI	Denotation-Demonstration-Interpretation
DHS	Demographic and Health Survey
DNA	Deoxyribonucleic acid
ERC	European Research Council
GRW	Ghirardi-Rimini-Weber
HD	Hypothetico-deductive
HDB	Hypothetico-deductive with bells on
INP	Integrated Nutrition Programme
NMA	No miracles argument
NOAA	National Oceanic and Atmospheric Administration
OTBF	Objectivity to be found
OWKI	Objectivity as we know it
PSA	Philosophy of Science Association

https://doi.org/10.1515/9783111019802-003

Matthias Egg

Introduction: An Empiricist View of Scientific Theories and Practices

The unique feature of the Lauener Prize for analytical philosophy is that it is a *twofold* prize: According to the Lauener Foundation's stated objective, the prize

> comprises, on the one hand, a lifetime achievement award (Lauener Prize for an Outstanding Oeuvre) and, on the other hand, an emerging talent award (Lauener Prize for Up-and-Coming Philosophers).

This twofold structure was reflected in a particularly attractive way in the line-up for the 2018 Lauener Symposium, which is at the origin of the present volume and brought together some of the leading philosophers of science of our time with some younger scholars who still have a large part of their philosophical journey before them. Being myself a member of this latter group, I found the symposium to be an immensely rewarding experience, providing me with a variety of insights into Bas van Fraassen's philosophical oeuvre, which had begun to shape debates in philosophy of science at a time when I was not even born. Since much of what was discussed at the symposium can now be read in the following chapters, I am confident that many readers will have a similar experience.

When a philosopher has influenced so many sub-fields over such a long period of time as van Fraassen, it is inevitable that most readers will first encounter his ideas in an indirect and somewhat anachronistic way; they will hear about them in the context of some current debate, usually without appreciating the historical background against which these ideas (and the very debates wherein they are now encountered) have taken shape. In my case, it was through one of his critics (and fellow Lauener symposiasts) that I first heard of van Fraassen and his opposition to scientific realism. As the following quotation makes clear, the context of this first encounter was the critique of an assumption—questioned by neither realists nor their opponents—about how theories are used in scientific practice to predict the phenomena we expect to observe:

> Van Fraassen's [1980] book takes a firm stand against realism. Sellars, I have mentioned, is a profound realist. But they have in common a surprising respect for theory. Both expect that theories will get the facts right about the observable phenomena around us. [...] This is not how I see good theories working. The observational consequences of the theory may be a rough match to what we suppose to be true, but they are generally not the best we can do. If we aim for descriptive adequacy, and do not care about the tidy organization of phe-

https://doi.org/10.1515/9783111019802-004

> nomena, we can write better phenomenological laws than those a theory can produce. (Cartwright 1983, 159–160)

What I did not realize for a long time was that van Fraassen's position had itself already resulted from profound transformations of how to view scientific theories, and that Cartwright's critique was, in a sense, just a continuation of his and other science-minded philosophers' struggle for a better match between our philosophical reflection on science and the actual practice of science. This struggle is a recurrent theme in many of the contributions to the present volume, and appreciating van Fraassen's role in it is not just of biographical or historical interest, but also leads to a better understanding of how philosophy of science became what it is today and how we might want to develop it in the future.

The essays collected in this volume therefore do at least two things: First, by highlighting some particular aspects of van Fraassen's work, they contribute to celebrating the outstanding philosophical creativity and poignancy of the 2018 Lauener laureate (more precisely: the senior one of the two laureates). Second, by engaging with these aspects in the context of current debates in philosophy of science, they reveal their continuing relevance and make them fruitful for addressing our present and future philosophical challenges. What this collection cannot achieve, of course, is to do full justice to all facets of van Fraassen's philosophy. This is made particularly clear in Steven French's *Laudatio*, which opens the collection and gives us some glimpses into both the depth and the width of van Fraassen's influence in the last half century of analytical philosophy.

The remaining chapters (with the exception of van Fraassen's *Reply to the Conference Participants*, which concludes the collection) are divided into three topical parts, each one focusing on one of the major themes of van Fraassen's work. The first part is concerned with the nature of scientific theories and their role in scientific representation, the second part looks at the persisting opposition between scientific realism and constructive empiricism, while the third part turns to the notion of epistemic stances and voluntarism. I will now briefly introduce each of these three topics and give an overview of the chapters making up the three parts.

1 Scientific Representation: Theories, Structures and Perspectives

If, as we saw in the above quotation, van Fraassen has "a surprising respect for theory", he stands in a long tradition of viewing theories as the primary vehicle of our scientific engagement with the world. The respect is nevertheless *surprising*, because van Fraassen is by no means an uncritical follower of that tradition. Quite the contrary, he has played a vital part in questioning prevalent views of what scientific theories are and how they contribute to our scientific understanding. His chapter on *The Semantic Approach, After 50 Years* retraces some of these developments.

The chapter starts by examining the shortcomings of the traditional account of theories ("the Received View"), in response to which van Fraassen and others developed what came to be known as "the semantic view" or "semantic approach". For the Received View, Euclid's geometry served as a paradigm for the general form of scientific theories: Briefly put, a theory can be viewed as a set of statements with an axiomatic basis and relations of logical deduction among them. One problem with this is that our actual scientific theories look very differently: Even if they can be cast in an axiomatic form (as in the case van Fraassen uses as an illustration here, namely quantum mechanics), the "axioms" of such theories serve a rather different role than in Euclidean geometry, and to analyze their relation to other statements in terms of first-order logic does very little to illuminate this role. As van Fraassen puts it: "the orientation toward formal logic, in the specific form it took in the Received View, was a distraction from understanding the sciences" (26). A second problem concerns the language in which the statements supposedly constituting a theory are formulated. The Received View incorporated the Logical Positivists' distinction between an "observational" and a "theoretical" vocabulary, thereby inheriting the well-known problems associated with that distinction. What van Fraassen highlights here is not so much the impossibility of drawing a clear boundary between the observational and the theoretical parts of language, but "the treacherous conviction that scientific language is in itself unintelligible" (27). Against this conviction, he argues that scientific language is part of natural language, so there is no need to reinterpret theoretical terms like "electron" or "gene" in terms of a (supposedly well-understood) observational vocabulary.

In response to these problems, the semantic approach sought "to direct philosophers' attention away from vocabulary and logical grammar to models and modelling" (27). This does not mean that formal logic becomes irrelevant for the analysis of theoretical discourse. Indeed, the chapter provides a detailed account of how different formal approaches to the semantics of theories and corresponding developments in logic crucially shaped the new semantic alternative to the Received View that emerged around 1970. It also describes the profound influence which the "quantum mechanics interpretation wars" had on this development. For readers not yet familiar with this part of van Fraassen's work, this presentation has the additional benefit of clarifying the connection of his approach to quantum mechanics (in particular, his 1991 book *Quantum Mechanics: An Empiri-*

cist View) with his more general take on scientific theories, as manifested in *The Scientific Image* (1980) and *Laws and Symmetry* (1989).

The chapter's presentation of the semantic approach and its history does not stop there, however. The approach's shift of emphasis from axiomatized systems of statements to sets of models proves equally fruitful in the context of less formal sciences, as van Fraassen shows by highlighting the central role played by model construction in Darwin's arguments for his theory of evolution. Another virtue of the semantic approach is that it promises to clarify the notion of empirical equivalence, which looked deeply puzzling from within the Received View. This notion is in turn intimately connected to empirical adequacy, the central concept of van Fraassen's constructive empiricism (to be discussed in Part II of this book). Finally, the way models are used in science also connects to van Fraassen's more recent work, especially his 2008 book *Scientific Representation: Paradoxes of Perspective*. For it is the *representational* function of models that lifts models out of the purely formal realm:

> Models are mathematical structures, but they are structures used for representation, and it is this function, of representing the empirical phenomena, that makes them scientific models. (71)

Steven French's chapter *Representation and Realism: On Being a Structuralist All the Way (Up and) Down* seamlessly continues these reflections on how mathematical structures can represent worldly phenomena. The title's reference to realism might suggest that the chapter could just as well have ended up in Part II of the book, but that would have amounted to (misleadingly) emphasizing the opposition between French and van Fraassen at the expense of their agreement. Indeed, the main topic of the chapter is how scientific representation can be understood in structuralist terms, regardless of whether one approaches this issue as a structural realist (like French) or as a structural empiricist (like van Fraassen). Starting from van Fraassen's insistence that nothing is a scientific representation unless it is *used* (by someone) as a representation, French addresses the worry that the very notion of scientific representation essentially involves a non-structural element, namely the *user* of the representation. In other words, any application of a theory to a real-life situation requires what van Fraassen calls an "act of self-location" in which the user identifies which part of the structure represented by the theory applies to her situation.

There is broad agreement between French and van Fraassen on how to respond to this worry, but their different outlooks nevertheless surface when French turns to a further challenge issued by van Fraassen, called the "Appearance from Reality criterion": A realist understanding of science requires that what we observe in measurements is (in some sense) produced by what the realist takes to be the unobservable reality (which, furthermore, is purely structural according to French's brand of realism). Here the discussion turns once more to quantum mechanics and its notorious measurement problem, which can be viewed as motivating a "methodological rejection" of the Appearance from Reality criterion. In the course of responding to this challenge, French helpfully distinguishes three different ways of spelling out the so-called Copenhagen approach to quantum mechanics and compares their implications for the Appearance from Reality challenge to each other and to the one suggested by the Everett interpretation. The upshot is that "the criterion can be met, and in an interestingly different way in each case" (99), depending on what exactly we take to be included in "the theory" as it is used for representing the target system.

The scare quotes in the last sentence indicate that we are now back to the topic of the previous chapter: What is a scientific theory? French has developed a highly original approach to this topic over the last few years, provocatively captured in the title of his recent book: *There Are No Such Things as Theories* (2020). According to this view, the word "theory" should not be taken as referring to a *thing*, but to "a range of practices associated with experimental work, as well as with modelling and theoretical manipulation" (102). This is reminiscent of van Fraassen's use-oriented view of scientific representation. It should therefore not come as a surprise that van Fraassen, in his reply to French, welcomes this move towards "a broadly pragmatic—or perhaps more accurately, phenomenological—approach to what is meant by our talk about scientific theories" (212), thus once again signalling a certain rapprochement between his empiricism and French's realism.

Another respect in which the notion of scientific representation proves its potential to foster a constructive dialogue between empiricism and realism is treated in Michela Massimi's chapter on *The Perspectival Nature of Scientific Representation.* Van Fraassen already emphasized that any representation *of* something is also a representation *as* something, for example when "the kinetic theory of heat represents a gas as an ensemble of molecules" (75). This means that any representation takes a certain perspective on what it represents, and Massimi examines different kinds of such perspectivity. One already familiar way in which representations become perspectival is by their user's self-location (see above). Massimi discusses this under the heading of *indexicality* in the last section of her chapter. The main part of her inquiry, however, is concerned with the perspectival nature of a representation's *content*, not only of its use. This is where the most pressing problems arise for anyone who seeks to reconcile what she calls *the representationalist assumption* with the so-called *perspectivalist assumption.* These two assumptions pull in opposite directions, in that the former asserts that scientific models (partially) represent relevant aspects of a given target system, while the latter states that they often do so in mutually incompatible or even inconsistent ways. In an attempt to do justice to both assumptions, Massimi works towards a position called *perspectival realism* (see also Massimi 2022).

There have been earlier attempts in that direction, and Massimi discusses some of their problems. She also draws an insightful connection to recent work by Catherine Elgin, who sharpens the perspectivalist assumption by claiming that scientific models often introduce "felicitous falsehoods" (e.g. Elgin 2017, 1), but nonetheless acknowledges models as "true enough", thus salvaging (to some extent) the representationalist assumption. But the most innovative aspect of Massimi's contribution is that she introduces a novel kind of perspectivity to the debate. In addition to the common idea that every representation is perspectival by representing its target *from* a specific vantage point, she draws attention to the directionality of a representation "*towards* a vanishing point that creates the perspectival illusion of a space extending beyond the boundaries of the represented scene" (114). It remains to be seen how much mileage the realist can get out of this (self-admittedly illusionary) aspect of perspectival representation, but it is certainly an aspect of the "perspective" metaphor that has so far been neglected and is worth exploring in search of a fuller understanding of how we gain scientific access to a complex reality by comparatively simple representations.

2 Constructive Empiricism and Other Foes of Scientific Realism

Few philosophical positions are as tightly bound to the name of a single philosopher as constructive empiricism is to the name of Bas van Fraassen. And since constructive empiricism has been one of the most prominent alternatives to scientific realism in the last four decades, many considerations of how science relates to reality are somehow connected to van Fraassen's work. This is reflected in the next three chapters of the volume, the first of which raises a novel objection to constructive empiricism, while the other two are dealing with more general issues in relation to scientific realism and its discontents.

Finnur Dellsén, winner of the 2018 Lauener Prize for Up-and-Coming Philosophers, does not spare his fellow Lauener laureate from criticism. His contribution with the title *Expanding the Empirical Realm: Constructive Empiricism and Augmented Observation* first of all provides the reader with a concise account of how van Fraassen disputes the scientific realist's view of the aim of science by contrasting it with the constructive empiricist view of that aim. (Very roughly: the aim of science is not truth as such, but only truth about the observable parts of the world, in other words, empirical adequacy.) Dellsén then goes on to review two familiar challenges for constructive empiricism: First, since empirical adequacy is advocated as a more modest aim than truth, why not become even more modest and only aim for "manifestational adequacy", defined as truth about the *observed* (as opposed to observ*able*) parts of the world? The second challenge stems from the variability of what is *observable*; different observers (or groups of observers) have different observational capacities, so what counts as empirically adequate differs from one epistemic community to the other.

Constructive empiricism has replies to these challenges. The first challenge is defused by noting that manifestationalism (the view that science should only aim for manifestational adequacy) has an absurd consequence which constructive empiricism does not have, namely that the aim of science could be served by choosing to make as few scientifically significant observations as possible. This shows that manifestationalism (in contrast to constructive empiricism) is too modest to make sense of scientific activity. In response to the second challenge, constructive empiricists maintain that relativizing the notion of empirical adequacy to an epistemic community poses no problem for their position.

Dellsén's innovative move now is to combine these replies and to turn them into a novel argument against constructive empiricism. By sketching various ways in which we could extend the boundaries of what is observable (*to us* as an epistemic community), he shows that it is to some extent up to us what empirical adequacy amounts to. And this implies that we could (in a parallel fashion to the manifestationalists mentioned above) serve the aim of science by avoiding, as best we can, any enhancement of our observational capacities. Dellsén concludes that constructive empiricism repeats the manifestationalist vice of proposing an aim of science that is too modest to account for what scientific communities usually strive for.

Among all the claims defended by the authors of this volume, this is probably the one that most directly opposes a position held by van Fraassen. The reader will therefore be forgiven for immediately turning to the end of the book for a peek into van Fraassen's response to it. Without spoiling the suspense, I can reveal that I found that response to be an exemplar of a simultaneously incisive and constructive reaction to criticism.

Milena Ivanova's chapter on *Aesthetic Virtues and Theory Acceptance* tackles a central aspect of how theories are judged when considerations of empirical adequacy are insufficient. Even empiricists acknowledge the importance of such considerations, which often involve aesthetic judgements, assessing, for example, a theory's elegance or beauty (or lack thereof). The point of contention between empiricists and scientific realists is, however, whether such aesthetic qualities have anything to do with a theory's likelihood of being true. Those who deny such a link from beauty to truth view the aesthetic virtues of theories as purely pragmatic, whereas many scientific realists have argued for an epistemic import of these virtues.

Ivanova discusses three types of realist arguments to that effect and shows how they fail to establish the sought-after link between aesthetic quality and likelihood of truth. The first type of argument is *ontological*, based on the idea "that a simple, elegant, unified or beautiful theory is more likely to be true simply because it reflects the simplicity, elegance, unity or beauty of nature" (151). This is unconvincing due to its reliance on the questionable a priori assumption that nature is simple, elegant, and so on. Some realists have therefore turned to a more *empirical* type of argument, using the historical record of scientific theories to argue for the reliability of inferences from aesthetic virtue to probable truth. Unfortunately, however, the historical record is rather ambiguous, as Ivanova points out by quoting Sabine Hossenfelder: "not only does the history of science thrive with beautiful theories that turned out to be wrong, but on the flipside we have the ugly ideas that turned out to be correct" (Hossenfelder 2018, 31). Thirdly, Ivanova considers the *no virtue coincidence argument*, recently put forward by Samuel Schindler. It starts from the observation (stressed in particular by Pierre Duhem and Thomas Kuhn) that scientists' aesthetic values and the way in which they influence theory assessment is subject to change in the course of history. Schindler takes this to support realism, because it would be a coincidence if a community of scientists—with all their different views on theoretical virtues—agreed on accepting a theory, unless that theory were true. In response, Ivanova points out that there is "substantial overlap between communities as to what aesthetic properties of theories are valued" (157) and that therefore, the alleged coincidence may not be as significant as required by Schindler's argument.

To this negative assessment of existing strategies to bring aesthetic considerations to bear on scientific realism, Ivanova adds a perspective from cognitive science, which continues the above-mentioned theme of bringing philosophy of science closer to actual scientific practice. Taking that practice seriously also means to acknowledge that scientists are non-ideal cognitive agents, whose behaviour can be investigated empirically. What these investigations suggest is that psychological factors like processing fluency or repeated exposure strongly influence our acceptance of hypotheses, to some extent independently of whether those hypotheses are likely to be true. By drawing attention to this growing body of research in cognitive psychology, Ivanova gives us one more reason to share van Fraassen's scepticism about the scientific realist's hope to identify true theories by looking at their extra-empirical virtues.

The final chapter in this second part of the book is Michael Friedman's *Realism and Instrumentalism: The Best of Both.* Unlike the two preceding chapters, this contribution is not primarily concerned with arguments either for or against scientific realism. Taking inspiration from a famous paper by Howard Stein, Friedman instead seeks to identify a point of contact between the two opposing camps, "whereby we accept "realism" but embrace what Stein calls "instrumentalism" at the same time" (166). (It does not matter much for our purpose that the non-realist camp is here called "instrumentalism" instead of "constructive empiricism"; its central feature is still that it attributes to science some other role than to inform us about the unobservable parts of the world.) The point of contact Friedman discusses is the method of *theory-mediated measurement*, which combines an instrumentalist and a realist view of science, as follows: Its instrumentalism consists in using theories as instruments to establish empirical access to some scientifically postulated quantity that is not directly measurable. At the same time, the method also lends itself to realism, by facilitating "inferences that are more secure than either inductive (curve fitting) or hypothetico-deductive arguments" (173). Friedman links this greater security to Kant's conception of causal necessity and goes on to show how this conception, although originally introduced in a context of Newtonian physics, can at least partly survive the Einsteinian revolution and be given sense within the context of general relativity.

The main focus of Friedman's chapter, however, is on Newton's usage of theory-mediated measurement in determining what was then called the "quantity of matter" of bodies in the solar system. The chapter provides a fascinating reconstruction of Newton's reasoning in Book 3 of the *Principia*, demonstrating how Newton transformed the existing notions of "quantity of matter" and "weight" to arrive at his conception of (inertial and gravitational) mass and to render this quantity empirically accessible. Readers who are less interested in this particular case study than in its implications for the epistemology of science may find it helpful to turn to van Fraassen's reply to Friedman, which discusses a simpler example and thereby reinforces Friedman's point. This alone suffices to prove that Friedman has succeeded in finding a point of contact between scientific realists and their opponents.

3 Epistemic Stances and Voluntarism

Although it is still a hotly debated topic (cf. Dellsén's chapter), the doctrine of constructive empiricism is by no means van Fraassen's final word on how empiricism is to be carried over into the twenty-first century. Indeed, his book *The Empirical Stance* (2002) has the potential to fundamentally transform not only our conception of empiricism, but of philosophical positions in general. There are two (closely interrelated) ideas underlying this transformation. First, van Fraassen encourages us to think of philosophical positions primarily as stances rather than doctrines. Second, he emphasizes the role of the will (hence "voluntarism") in our adopting of stances and, consequently, of beliefs as well. The two chapters in this final part of the book explore some ramifications of these ideas.

The first one is my own contribution to this volume, discussing *Stances and Doctrines in Scientific Metaphysics.* It assesses the impact of van Fraassen's legacy on an influential approach in contemporary analytic metaphysics, namely James Ladyman and Don Ross's version of scientific metaphysics. These authors share with van Fraassen a strong aversion to metaphysical doctrines that purport to pursue the same standards of objectivity as the sciences, but are actually paying very little attention to the results and practices of real science. This aversion, however, does not lead Ladyman and Ross to abandon the metaphysical project (as van Fraassen would advocate), but to reconfigure it in a way that is no longer subject to such shortcomings. One part of this reconfiguration consists in purging metaphysics of its doctrinal character, by going along with van Fraassen's transition from philosophical doctrines to stances. The result is what Ladyman and Ross call *weak* metaphysics. My chapter critically reviews the alleged virtues of this transition and argues that weak metaphysics is still almost as doctrinal as the older kind of metaphysics it is supposed to replace.

Nevertheless, the focus on stances instead of doctrines may prove to be illuminating. It may, for example, draw our attention away from the possibly unanswerable question of who is wrong or right in the scientific realism debate (see Part II). Unlike doctrines, stances aren't true or false, so we may simply have to accept that the realist and the empiricist are equally within their rights to approach the issue from their respective stance. In other words, there may be no basis for judging either the metaphysical (realist) or the empirical stance as superior to the other. This is how Anjan Chakravartty has recently portrayed the situation, advocating *aphasia* (speechlessness, suspension of judgement) as the proper response to the question of superiority among rational stances. In the second part of my chapter, I argue that this conclusion is premature. My argument against it builds on Chakravartty's own characterization of what it takes for a stance to count as rational, which crucially involves a rather demanding notion of internal coherence. If we take this requirement seriously, we may find that some stances are superior to others in the sense that they display a higher degree of internal coherence. This may ultimately enable us to rank stances according to a stance-independent measure of rationality. I take a step in that direction by pointing to a way in which the so-called scientistic stance (on which Ladyman and Ross base their scientific metaphysics) may fail to be internally coherent. In his reply to my chapter, van Fraassen adds some further points to this negative assessment, but of course, this still leaves room for constructive work on a more coherent stance, especially for those who are (like myself) not quite comfortable with van Fraassen's metaphysics-free empirical stance.

Chakravartty's *aphasia* is interestingly connected to Nancy Cartwright's chapter *Evidence, Relevance and Warrant: In Defence of Voluntarism.* We may not only be at a loss of words when comparing different stances *from the outside*, but also when asked to justify *our own* adoption of a certain stance. "That's just how I see it" may ultimately be the bedrock of our justificatory endeavours. Cartwright coins a picturesque term for this way of ending a justificatory chain: *plumping.* In contrast to the discussion in the previous chapter, she emphasizes the need to plump not only at the abstract level of philosophical stances, but at the everyday level of warranting even the most mundane empirical claims: "plumping is ubiquitous" (202).

A part of the problem (but, as we shall see, only its simpler part) is familiar from what in epistemology is known as Agrippa's (or Münchhausen's) trilemma, and similarly from our childhood (or parenthood) experience that any answer to a why-question can be followed by yet another why-question. Put more technically, if a hypothesis h is justified by providing some evidence e, the claim e in turn requires justification by further evidence, and so on. The only way to avoid an infinite regress or circularity is to plump at some point. This is already problematic, but the harder part of the problem, as Cartwright points out, does not concern the question whether e is *true*, but whether it is *relevant* (for supporting h): "Warranting an evidence claim e is a doddle compared with warranting that e *is* evidence" (194). This part of the problem may be more or less conspicuous, depending on one's favourite theory of evidence, but it is there anyway. To demonstrate this, Cartwright discusses it in the context of two different such theories. The first one is "the argument theory of evidence", to which she herself subscribes. It construes the notion of evidence as a three-place relation between e, h and A: "e is evidence for h relative to a good argument A for h" (198). Since A only counts as a good argument if its premises are warranted, these premises require additional evidence, which in turn requires a good argument, and so forth. As a second theoretical context, Cartwright considers the famous hypothetico-deductive (HD) account of evidence, according to which e is evidence for h if it can be deduced from h and is true. Since "being evidence for" is here construed as merely a two-place relation between e and h, the need to plump for a certain take on when exactly e *is* evidence for h may easily be overlooked. But it again becomes obvious as soon as one looks at the details of the HD account (involving, as it must, auxiliary hypotheses to deduce e from h and additional requirements for some fact e deduced from h to count as supporting h).

These are not just idle philosophical quarrels. If our policies are to be "evidence-based", then what counts as evidence literally becomes a question of life and death, as Cartwright's examples of the Bangladesh Integrated Nutrition Programme and the Vajont dam disaster drastically illustrate. Cartwright thereby joins van Fraassen in highlighting an ethical—and in some sense existentialist—dimension of the empirical stance: The empirical evidence on which our beliefs (and actions) ought to rely is not simply *given* to us. Instead, we are responsible for our beliefs by plumping for a certain commitment concerning what we *take* to be evidence.

References

Cartwright, N. 1983. *How the Laws of Physics Lie.* Oxford: Clarendon Press.
Elgin, C. Z. 2017. *True Enough.* Cambridge, MA: MIT Press.
French, S. 2020. *There Are No Such Things as Theories.* Oxford: Oxford University Press.
Hossenfelder, S. 2018. *Lost in Math: How Beauty Lead Physicists Astray.* Cham: Springer International Publishing.
Massimi, M. 2022. *Perspectival Realism.* Oxford: Oxford University Press.
van Fraassen, B. C. 1980. *The Scientific Image.* Oxford: Clarendon Press.
van Fraassen, B. C. 1989. *Laws and Symmetry.* Oxford: Clarendon Press.
van Fraassen, B. C. 1991. *Quantum Mechanics: An Empiricist View.* Oxford: Clarendon Press.
van Fraassen, B. C. 2002. *The Empirical Stance.* New Haven, CT: Yale University Press.
van Fraassen, B. C. 2008. *Scientific Representation: Paradoxes of Perspective.* Oxford: Clarendon Press.

Steven French

Laudatio: Professor Bas van Fraassen

I'd like to begin by thanking Michael Frauchiger and the other members of the Council of the Lauener Foundation for inviting me and giving me this opportunity to offer some personal reflections on Bas's philosophical work.

In the opening lines of a paper published in the journal *Philosophy of Science* at the end of the 1970s, entitled "Scientific Realism: The New Debates", Edward Mackinnon writes:

> In the last few years the traditional debate between realist and instrumentalist interpretations of scientific theories has suffered a sea change into something rich and strange. (Mackinnon 1979, 501)

Leading that shift, Mackinnon lists Mary Hesse, Joseph Sneed, Wolfgang Stegmüller, Richard Rorty and Bas. As stellar as that particular line-up is, it is Bas's work that has perhaps had the most lasting and extensive impact across so much of the philosophy of science. Just how rich and yes, strange, at least to certain realists, that transformation of the debate was, became clear the next year with the publication of *The Scientific Image* (van Fraassen 1980), its chapters sketched out round the campfire as Bas travelled through southern Europe and north Africa. It's a cliché to say that the term "classic" has become a cliché but *The Scientific Image* is surely one of the most significant and transformative books in the philosophy of science of the last 100 years. It has been cited thousands of times, remains a core text on many reading lists, and prompted an outstanding collection of responses in Paul Churchland and Clifford Hooker's *Images of Science* (Churchland and Hooker 1985), with replies from Bas, as well as numerous reviews and commentaries, including some notable and well-known examples from colleagues here today (see Friedman 1982). But I'm willing to bet that few of you are familiar with the review posted by a certain Adam, on the website Goodreads, who writes that *The Scientific Image* is "profoundly badass"! For those unfamiliar with the vernacular, the Merriam-Webster definition of "badass" is that "of formidable strength and skill"—which aptly describes the book, rehabilitating, as it does, the anti-realist project in the philosophy of science and thereby laying down the basis of a new form of empiricist philosophy more generally.

The trajectory of Bas's constructive empiricism from that point has been much discussed, through to *The Empirical Stance* (Yale University Press, 2002), with, as the title suggests, its advocacy of a novel and provocative stance-oriented perspective, and beyond, to what Michela Massimi has called the "paradigmatic expression

https://doi.org/10.1515/9783111019802-005

of [...][a][...] new trend in philosophy of science" (Massimi 2009, 323), namely the book *Scientific Representation* (Oxford University Press, 2008), where Bas's "empiricist structuralism" blooms in full. But if I could shift the emphasis a little, what was also significant about *The Scientific Image* and the further elucidation of its core claims by Bas in his replies to the critics, at least for those of us who came into philosophy of science with a scientific background, was what Mackinnon highlighted as the "focus on functioning science", rather than ideal reconstructions (MacKinnon 1979, 502). This is manifested in multiple ways but I'd like to focus on two in particular. First of all, there is Bas's central point that the "axioms" that one finds in textbooks on quantum mechanics, say, "do not look very much like what a logician expects axioms to look like" (van Fraassen 1980, 65), together with his proposal that they should be seen, instead, as specifying families of structures. This introduction to what is now generally known as the Semantic Approach came as a revelation, not just to me but to many others, of course, and as Bas himself insisted, many of the familiar issues and problems in the philosophy of science were transformed in light of it.

Both this shift in focus and the overall attitude are also apparent in his subsequent book, *Laws and Symmetry* (1989). The majority of the reviews of this work seem to concentrate on only the first of Bas's two objectives: to demonstrate the failure of some of the more well-known philosophical accounts of laws of nature and to undermine the epistemological principles behind arguments for the reality of such laws; see for example the symposium on the book in *Philosophy and Phenomenological Research* from 1993 (to which Nancy Cartwright contributed; Cartwright 1993). But as Bas makes clear, in his response, what was closest to his heart was the constructive aspect, according to which laws should be seen as a vestigial concept in contemporary physics, where arguments from symmetry and continuity dominate. Again, this insertion of such core concepts from current physics directly into discussions in general philosophy of science was hugely significant and prescient. Framed in terms of the Semantic Approach, they were presented as elucidating the structure of the relevant models and in this work Bas foreshadows the later spotlighting of symmetry that we find in Katherine Brading and Elena Castellani's collection *Symmetries in Physics* (Brading and Castellani 2003). Bas, of course, also contributed to this volume in a piece with Jenann Ismael, where they explore the role of symmetry as a guide to "surplus" theoretical structure. Here they argue that methods for removing this superfluous structure lie at the very heart of theorizing in modern physics, where, as they put it

> [...] one starts with the representations and works one's way towards ideas about the intrinsic character of their common object by a kind of triangulation. (van Fraassen and Ismael 2003, 390)

Thus the ontologies of our most fundamental theories are guided not by physical intuition, nor are they shaped by philosophical prejudice but, Bas and Jennan argue, are led by this ideal of a kind of formal simplicity (van Fraassen and Ismael 2003, 392).

This "triangulation" underpins that other significant feature of Bas's earlier work that particularly resonated with me and many others, namely his comment, in his "Replies to Critics" from *Images of Science*, that the reason why metaphysics seems so passé today, is that it has been surpassed by theoretical science when it comes to intellectually challenging perplexities and paradoxes: "Do the concepts of the Trinity, the soul, haecceity, universals, prime matter, and potentiality baffle you? They pale beside the unimaginable otherness of closed space-times, event horizons, EPR correlations, and bootstrap models" (van Fraassen 1985, 258). Similar sentiments have since been expressed, albeit perhaps more brutally, by James Ladyman and Don Ross in *Everything Must Go* (Ladyman and Ross 2007), and although you might disagree, as Kerry McKenzie and I do, with both Bas and James, when it comes to the potential of so-called "analytic" metaphysics for playing a role in the philosophy of science, it has to be recognized that that earlier dismissal also raised a major challenge for the realist, particularly in the context of issues, some more or less well-known, arising from quantum mechanics.

Although Bas's 1991 spin-off from *Laws and Symmetry*, namely *Quantum Mechanics: An Empiricist View* (Oxford University Press, 1991), is perhaps more well-known for the lucid presentation of his "Copenhagen" variant of the modal interpretation, drawing on his ground-breaking work from the early 1970s, it is the defence of "semantic universalism" that may well better stand the test of time. Anticipating the recent work of Shamik Dasgupta (2009) and Simon Saunders (2018), this is centred on the claim that "all factual description can be completely given in entirely general propositions" (van Fraassen 1991, 465), and it imposes on our representations a fundamental constraint, namely Permutation Invariance. This maps onto one of the principles that helped usher in the reign of symmetry in modern physics and which, as manifested in quantum statistics, led the likes of Born and Heisenberg to declare, as early as 1926, that quantum particles must be regarded as non-individuals, in some sense. In what sense? Mary Hesse, and before her, Erwin Schrödinger, compared such particles to pounds or Swiss francs in a bank account; my old Professor, Heinz Post, likened them to the Cheshire cat from *Alice's Adventures in Wonderland*; and Maria Luisa Dalla Chiara, Newton da Costa and Décio Krause, together with their students and collaborators, have developed non-standard forms of logic and set theory to accommodate them (see for example Krause 1992). But as Bas shows, this particular metaphysical option is not forced on us by quantum statistics itself—the models required by semantic universalism can be described equally in terms of non-individuals or individuals. Thus we have here a kind of "metaphysical underdetermination" to add to the more familiar forms (van Fraassen 1991, 481), and James Ladyman and I understood the concluding section of Bas's book, entitled "Good-bye to Metaphysics", as setting out a further challenge to the realist, in addition to the more well-known admonitions about use of Inference to the Best Explanation and the argument of the Best of a Bad Lot: if she cannot even say whether an electron, say, is an individual or not, with the alternative metaphysical and formal packages that these differing assertions must come wrapped in, then her realism is surely "ersatz" (Ladyman 1998). I have to confess that I remain baffled by the apparent reluctance of the realist community in general to respond to Bas's challenge. John Worrall, for example, waves a hand in the direction of quantum physics in his groundbreaking paper that re-introduces structural realism (Worrall 1989) but declines to ride that particular wave. And to this day, although realist-inclined *interpretations* of quantum mechanics abound, consideration of the epistemological implications for realism more generally remains scant, as my colleague Juha Saatsi has lamented (2019).

Bas's own response, of course, was to press the adoption of his empiricist stance; mine and James was also to pull in our metaphysical horns, but not quite so far, urging instead that the realist drop her commitment to an object-oriented metaphysics and adopt a structuralist one instead. The differences between these stances have been extensively explored, as have their similarities and one point on which Bas and I certainly agree concerns the prospects of a "pure" structuralist ontology. As he notes in *Quantum Mechanics: An Empiricist View,* semantic universalism leaves something out: in order to apply science we need to refer to specific things in the world and so we need something that goes beyond theory (van Fraassen 1991, 466). What we need is not just indexical language but also the knowledge how to *use* it, so that we can situate ourselves appropriately with respect to our representations (and so, for example, Putnam's Paradox and Newman's objection to structuralism are defused; van Fraassen 2008, 232–235). This, for Bas, is where pragmatics enters and as he says at the conclusion of the chapter on "empiricist structuralism" in *Scientific Representation* (2008), "[t]his implies no relevant incompleteness in the theory or model itself" (van Fraassen 2008, 261), just as a map of the Swiss railway system is not incomplete because it comes without a label that says "you are here".

Thus, as he suggests in response to James Ladyman's critical review of the above book, "[p]erhaps, it could be seen as a metaphysical discovery, that absolutely pure structuralism is impossible?" (van Fraassen 2011, 439). But of course where differences may arise between us, and among structuralists of various stripes in general, is how the nature of the impurity is understood. So, consider Weyl's famous claim, made in the context of space-time theory and cited by Bas, that coordinate systems should be understood as "the unavoidable residuum of the ego's annihilation". Weyl, as we all know and as Tom Ryckman has examined in some detail (see van Fraassen 2007), adopted a *phenomenological* stance (not in the sense in which this term is used in physics today but in the Husserlian sense) and from this perspective the insertion of the ego into such considerations does not represent the blemish of some "impurity" but rather has to be seen as a consequence of the conditions for doing science, or indeed, acquiring empirical knowledge, at all! For Bas, similarly, the indexical is essential (van Fraassen 2008, 3; see also van Fraassen 2006) and to use the models that physics gives us requires that we locate ourselves with respect to them and hence results in the insertion of our own perspective (van Fraassen 2011, 436). The "pragmatic tautology" that is the identification of the claims "that a theory is adequate to the phenomena" and "that a theory is adequate to the phenomena as represented (by us)" is, as the name suggests, contingent but undeniable (van Fraassen 2008, 259). One might worry that such an appeal to the conditions of use of our models is "at odds with the idea of objective enquiry" (Ladyman 2011, 417), or whether acts of representation induce pragmatic commitments in the way that Bas's argument requires (Nguyen 2016) but given its unavoidability or essentiality within the stance adopted, such a move can hardly be viewed as a blot on the structuralist landscape (but then, I would suggest, by the same token, neither can the realist's appeal to the nomological character of the world-as-structure, that James and I emphasize; Ladyman 2011; French 2014).

There is a more general and, perhaps, more profound, point here, as Bas writes in *The Empirical Stance:*

> Science, whether understood with the scientific realist or with the empiricist, provides us with a world picture. Accepted in either sense, it can be our entire world picture. But if it is, we ourselves don't seem to fit into our own world picture. (van Fraassen 2002, 189)

And he goes on to argue that no objectifying inquiry, such as that of science, can reveal what persons are or who the persons are among things of the world (van Fraassen 2002, 191). This is a theme that Bas also pursues in a paper that he lists as one of his favourites, "The Transcendence of the Ego (The Non-Existent Knight)" (van Fraassen 2004), in which he nicely draws on Italo Calvino's novel from 1959 (as given in the parentheses above) to argue that "I am not a thing among things" (van Fraassen 2004, 453); that is, not a physical object, nor a mental substance nor an abstract entity. Instead, Bas replaces such identifications with an explicitly relational conception in which "I" *manifest* myself in nature (and in society), in different respects and to different degrees. (Likewise, as part of her attempt to remove the need for object talk in metaphysics Laurie Paul substitutes "manifestation" for "instantiation" in the case of properties and so the latter can be admitted into one's framework without the necessity for things that "possess" them (French 2014, 192–193).) As a result of such a manoeuvre, certain problems evaporate, such as, in the case of the Self, Bas argues, its (un)observability, its supervenience on the physical and the emergence of consciousness (van Fraassen 2004, 470–476). And perhaps, with the recognition of such as "pseudo-problems", philosophy itself may change, to become what it should be, namely, as Bas hopes in *The Empirical Stance*, "an authentic, engaged project in the world" (van Fraassen 2002, 195).

What I've sketched here are just some of the anchors of the ascent through Bas's oeuvre—running between them are multiple strands and lines of thought, from his initial papers on free logic, to his work on quantum logic and the logic of conditional obligation—and here we should also acknowledge his editorships of the *Journal of Philosophical Logic* and the *Journal of Symbolic Logic*—through the probability of conditionals to his recent work—published in the *British Journal for the Philosophy of Science*—on the general constraints that any policy on conditionalization should follow, and from Aristotle on tragedy and science to, of course, fundamental issues of interpretation in science and art—that last in a paper that, to get personal again, I continue to enjoy discussing with our students at Leeds (van Fraassen and Stigman 1993)—through—but hopefully not finally!—to his more recent work with Isabelle Peschard on the experimental side of modelling (Peschard and van Fraassen 2018).

It is of course impossible to do justice to over 50 years of ground-breaking philosophical research in a presentation such as this. But perhaps what we can say, in summary, is that Bas *manifests* himself, not only through his body and his actions, his climbing and hiking, his relationships with friends and family and colleagues, but also through this incredibly wide-ranging body of work, which is rich indeed and, perhaps, sometimes "strange" but also erudite, thought provoking and innovative and is now recognized, rightfully, as truly outstanding with this prestigious award.

References

Brading, K. and Castellani, E. (eds.). 2003. *Symmetries in Physics: Philosophical Reflections.* Cambridge: Cambridge University Press.

Cartwright, N. 1993. In Defence of 'This Worldly' Causality: Comments on van Fraassen's Laws and Symmetry. *Philosophy and Phenomenological Research* 53: 423–429.

Churchland, P. and Hooker, C. (eds.). 1985. *Images of Science: Essays on Realism and Empiricism.* Chicago and London: University of Chicago Press.

Dasgupta, S. 2009. Individuals: An Essay in Revisionary Metaphysics. *Philosophical Studies* 145: 35–67.

French, S. 2014. *The Structure of the World: Its Metaphysics and Representation.* Oxford: Oxford University Press.

French, S. 2016. Replies to My Critics. *Metascience* 25: 189–196.

Friedman, M. 1982. Review of *The Scientific Image. Journal of Philosophy* 79: 274–283.

Hardy, L. 2013. *Nature's Suit: Husserl's Phenomenological Philosophy of the Physical Sciences.* Ohio: Ohio University Press.

Ismael, J. and van Fraassen, B. 2003. Symmetry as a Guide to Superfluous Theoretical Structure. In: *Symmetries in Physics: Philosophical Reflections*, eds. K. Brading and E. Castellani, 371–392. Cambridge: Cambridge University Press.

Krause, D. 1992. On a Quasi-Set Theory. *Notre Dame Journal of Formal Logic* 33: 402–411.

Ladyman, J. 1998. What Is Structural Realism? *Studies in History and Philosophy of Science* 29: 409–424.

Ladyman, J. 2011. Review of Scientific Representation: Paradoxes of Perspective. *Metascience* 20: 417–442.

Ladyman, J. and Ross, D. 2007. *Everything Must Go.* Oxford: Oxford University Press.

Mackinnon, E. 1979. Scientific Realism: The New Debates. *Philosophy of Science* 46: 501–532.

Massimi, M. 2009. Review of Scientific Representation "Paradoxes of Perspective". *International Studies in the Philosophy of Science* 23: 323–327.

Nguyen, J. 2016. On the Pragmatic Equivalence between Representing Data and Phenomena. *Philosophy of Science* 83: 171–191.

Peschard, I. and van Fraassen, B. 2018. *The Experimental Side of Modelling.* Minneapolis: University of Minnesota Press.

Saatsi, J. 2019. Scientific Realism Meets Metaphysics of Quantum Mechanics. In: *Philosophers Look at Quantum Mechanics*, ed. A. Cordero, 141–162. Cham: Springer.

Saunders, S. 2018. The World Is the Totality of Quantities, Not of Things. *Presidential Address at the 2018 Meeting of the British Society for the Philosophy of Science.* https://www.thebsps.org/bsps-videos-4/.

van Fraassen, B. 1980. *The Scientific Image.* Oxford: Oxford University Press.

van Fraassen, B. 1985. Bas van Fraassen Replies to His Critics. In: *Images of Science*, eds. P. Churchland and C. Hooker, 244–308. Chicago: University of Chicago Press.

van Fraassen, B. 1989. *Laws and Symmetry.* Oxford: Oxford University Press.

van Fraassen, B. 1991. *Quantum Mechanics: An Empiricist View.* Oxford: Oxford University Press.

van Fraassen, B. 2004. Transcendence of the Ego (The Non-Existent Knight). *Ratio* 17: 453–477.

van Fraassen, B. 2006. Weyl's paradox: the distance between structure and perspective. In: *Bas C. van Fraassen: The Fortunes of Empiricism*, eds. A. Berg-Hildebrand and C. Suhm, 13–34. Frankfurt am Main: Ontos.

van Fraassen, B. 2007. "The Constitutive A Priori". Review of The Reign of Relativity. *Metascience* 16: 407–419.

van Fraassen, B. 2008. *Scientific Representation: Paradoxes of Perspective.* Oxford: Oxford University Press.

van Fraassen, B. 2011. Author's Response. *Metascience* 20: 417–442.

van Fraassen, B. and Stigman, J. 1993. Interpretation in Science and in the Arts. In: *Realism and Representation*, ed. G. Levine, 73–99. Madison, WI: University of Wisconsin Press.

Wilson, J. 2012. Fundamental Determinables. *Philosophers Imprint* 12(4): 1–17. http://hdl.handle.net/2027/spo.3521354.0012.004 (last accessed: 3. 8. 2022).

Worrall, J. 1989. The Best of Both Worlds. *Dialectica* 43: 99–124.

Part I: **Scientific Representation: Theories, Structures and Perspectives**

Bas C. van Fraassen

The Semantic Approach, After 50 Years

Abstract: The 1960s saw many revolutions, worldwide, and some of that epoch's revolutionary spirit manifested itself in philosophy of science, with strong reactions against the dominant "received view" of Logical Positivism. Scientific realism emerged to dispute ontology, Kuhn single-handedly turned our eyes back to history of science, and the semantic approach replaced the methodological framework for philosophers of science. The Logical Positivist revolution had just about reached age 50 at the time; today the semantic approach is about 50 years old as well, and faced with a plethora of new developments. I will discuss how the semantic approach was born, and its fortunes and misfortunes, from its beginnings to approximately 1990, with just a brief look at what has been happening since then.

1 Introduction

The semantic approach has had an exciting history deeply entangled with other twentieth-century philosophical currents which led up to the philosophy of science of today. Much of its early history was already chronicled by Frederick Suppe in several publications (Suppe 1974, 1989, 2000). I will add my own perspective, and concentrate especially on certain controversies in which the semantic approach played a distinctive role. Although I will focus on the years before 1990 I will add a discussion of more recent developments that I take to belong to, or be a proper sequel to, the semantic approach. This will be partial, fragmentary, if only because it is hard to see the wood when in the midst of the trees.

I was a participant in this history, so my account will not be impartial. Episodes on which some cast a jaundiced eye I may be seeing through rose-coloured glasses. On the origins of the semantic approach I am in agreement with Frederick Suppe (whose 1967 dissertation had developed a full-blown version of the new semantic approach before we started using that name), but there certainly remain open questions.[1] Suppe and I shared a good deal of reflection on just where we had gotten our ideas, and those early conversations may have given me a somewhat narrow view, in ways that I can't hope to detect introspectively with any accuracy.

1 There are contrary claims, such as "The semantic view was first introduced as an interpretative method by E. W. Adams" (Collier 1992, 288).

https://doi.org/10.1515/9783111019802-006

2 Death of the Received View

About half a century ago, the Logical Positivist's revolutionary approach to science was itself already approximately 50 years old. At that time their account of the structure of a scientific theory was dominant in the literature. Hilary Putnam could appropriately call it the *Received View* in his critical "What Theories Are Not" (Putnam 1962), signalling both by this nomenclature and by his title that a new revolution was brewing.

2.1 A Death (?) and a Birth (?)

Can the date of that second revolution in philosophy of science be fixed? Not exactly, I think. Frederick Suppe did offer a precise date:

> The *Received View of Theories* was the epistemic heart of Logical Positivism. Twelve hundred persons were in the audience the night it died. It was March 26, 1969—opening night of the Illinois Symposium on the Structure of Scientific Theories [which Suppe had organized]. (Suppe 2000, 102)

Perhaps so. We cannot add that on this very same night the *semantic approach* was born.[2] Not a single contribution to that symposium appears to introduce it, except by negative implication so to speak. The new semantic approach was only presented in Suppe's "Introduction to the Proceedings" of that symposium, published five years later (Suppe 1974, 221–230). Yet I think that the birth of the semantic approach can be located in the sixties, although vaguely, with a beginning prior to that fateful night, and with a clear pre-history in the several decades before that.[3]

I won't blow the semantic approach's horn, or tout its pervasive presence in philosophy of science in the past half century. I don't need to, for that was done very well by some of its recent critics:

> Within a few short decades, the semantic view has established itself as the new orthodoxy. According to Roman Frigg (2006, 51), "Over the last four decades the semantic view of theories has become the orthodox view on models and theories". One only has to glance at recent writ-

---

2 I will use the name "semantic approach" as well as Suppe's "semantic view", because much more than views on theory structure was involved in this development.

3 Frederick Suppe also gives a quite detailed historical account in a later book (Suppe 1989, 3–20). It has more details than I will be able to give here, about how Suppe and I interacted, and about the routes by which we arrived at the semantic approach, as well as about its later development.

> ings on the philosophy of science to verify Frigg's claim: the semantic view is now assumed as the default explication of the notion of a scientific theory. (Halvorson 2012, 183)

This was Halvorson's preamble to a severe critique, which I will address below. Halvorson goes on to document, with examples and with some dismay, how the semantic approach transformed the scientific realism debates, the philosophy of biology, philosophy of psychology, and philosophy of physics. Critics have a stake in the importance of their target, as an innocent bystander would not. So it may be more telling to mention Thomas Kuhn's remark, after he had asked Elisabeth Lloyd to explain the semantic approach:

> With respect to the semantic view of theories, my position resembles that of M. Jourdain, Moliere's *bourgeois gentilhomme*, who discovered in middle years that he'd been speaking prose all his life. (Kuhn 1993, 3)

2.2 The Received View—and What Was That?

Before we look at the development of the semantic approach, just what was the Received View that died that night? Hilary Putnam, who introduced the name, was himself present in the Illinois conference. His contribution was a detailed continuation of his critique, focusing on explanation, confirmation, and corroboration (Putnam 1974). Putnam gave, so to speak, the counterpoint to Hempel's paper which opened the conference, though Hempel called that view of theories, surely with equal warrant at that time, "the Standard View". Neither their papers, nor the other contributions on that original occasion presented a clear and concise overall account of that view. That too was supplied a little later, by Steven Toulmin.

2.2.1 Toulmin's Take on the Received View

In a Postscript to the Proceedings of the Illinois Conference, Toulmin worked mightily to explain just what the Received View had been, and what he took to be now roundly rejected therein, to clear the decks for a new approach to scientific theories.

A crucial clue to the Logical Positivists' view had been the continuing influence of Euclid's paradigm for theory-form, as extrapolated in the seventeenth century to the idea of a *mathesis universalis:*

> In its fully developed form the traditional (or Viennese) philosophy of science incorporated into itself [...] the symbolism of mathematical logic and the methodological program of the unified science movement. At its most grandiose, the ambition was to establish epistemological foundations [...] for positive scientific knowledge in its entirety. What [...] Peano and Russell had done by founding pure mathematics in formal logic [...] the advocates of unified science hoped to do for the whole of natural science. By adding further primitive terms, postulates, and correspondence rules, they hoped to incorporate all genuine branches of science into a single axiomatic edifice [...]. (Toulmin 1974, 602)[4]

Briefly, the Received View of theory structure was therefore that a scientific theory can be modelled by what logicians call a theory or deductive system: an axiomatized set of sentences in a language designed for that purpose. Toulmin mentions only briefly what was perhaps the most contentious aspect of this view: that the vocabulary of that language is divided into *observational terms* and *theoretical terms*, and that the latter derive their meaning and reference from "correspondence rules" that link them to the former.

The objection was not that it is in general impossible to have grand axiomatic edifices. The objection was rather that such a program leads us up the garden path: it leads us from close attention to the scientific theory itself to problems to be solved in formal semantics and logic, at a different level of abstraction. Nor was the objection a rejection of the value or use of formal methods. Rather, the orientation toward formal logic, in the specific form it took in the Received View, was a distraction from understanding the sciences.

Let me give an example, for apart from this motivation, the dispute is well-nigh empty. Think about the basic principles of quantum mechanics, such as

- States of a physical system ... are represented by the vectors ψ in a linear space ...
- Attributes of the system ... are represented by linear self-adjoint operators Q on the space of vectors ...

as they appear typically in a textbook (e.g. Peebles 1992, 104–105; 188–189).

A language in which the whole of the theory is formulated would have to have sentences which say precisely that.

> Could it be a language of first-order logic? Certainly! You could give their symbolization as an exercise to your logic class: choose "A" for "attribute", "S" for "self-adjoint", ..., and be sure to get the quantifiers in the right order...!

---

4 This revival of Leibniz' *mathesis universalis* is what Fred Suppe had already taken on as a target for critique in his dissertation (Suppe 1967; cf. Suppe 1974, 221–230). It was equally what I saw as rejected in the lesson that Evert Beth drew from Birkhoff and von Neumann, as I will detail below.

But compare this with the exercises students get in the physics class, such as:

> Find the energy levels in a harmonic oscillator.

The physics student constructs a model guided by those principles. For example, a harmonic oscillator the textbook defines straightforwardly as a system in which the total energy is a certain function of position and momentum. But now the student has to apply those principles, so that energy, position, momentum are represented by linear operators *H, P, Q*. The result is that his problem translates into finding the eigenvalues of the operator ½ $(P^2 + Q^2)$ subject to the condition that $(QP - PQ) = ih$, which turn out to form a discrete series of fractions. As a result, the lowest energy level cannot be zero, and gain or loss of energy is by discrete amounts, quite contrary to what a classical model would show.

Obviously, the logic student learns nothing about quantum physics and the physics student nothing about first-order logic. What would be the point of some gargantuan formalizing project?

If we want to understand the science, we will do better to concentrate on what it is the physics student learns, namely, how the equations, or rather their solutions, can represent the phenomena. That was our point. It certainly does not imply banishing language from scientific theorizing—a trivially absurd idea—but it requires us to put language in its proper place, and to direct philosophers' attention away from vocabulary and logical grammar to models and modelling.[5]

2.2.2 The Treacherous Conviction That Scientific Language Is in Itself Unintelligible

Toulmin's description of our discontent, and of our rejection of the earlier tradition, leaves out one crucial part: the question of how to place the language of science within the philosophical compass.

The Received View is more or less what we took (whether correctly or incorrectly) to be the main content of the Logical Positivist approach of Carnap and his cohort. And that had as its most striking feature the conviction that *theoretical language stands direly in need of interpretation.* Theoretical vocabulary is, in itself, unintelligible! In this respect theoretical language purportedly contrasted with the

---

5 Thus, as I will elaborate below, both Fred Suppe's (1967) and my (1970) were in good part essays on the description of scientific models in certain kinds of languages, though not in the form to be associated with the earlier accounts of e.g. Carnap, Hempel, or Nagel.

language in which observational evidence is reported, which we were supposed all to understand. (Pierre Duhem's insights were apparently forgotten for some time.) So the interpretation had to be effected through, if possible, a reduction, and if that was not feasible, a web of links of the theoretical terms to the observation idiom.

That view of scientific language was alien to us, largely due to the salient and vocal presence of the Minnesota scientific realists like Grover Maxwell and my teacher Wilfrid Sellars, as well as the newly famous Paul Feyerabend (1957), Thomas Kuhn (1962), and my colleague at Yale, Norwood Russell Hanson (1958).[6] These had, through their provocative, inspired, and passionate advocacy, instilled in us the contrary conviction that there is no line to be drawn between the "observational" and "theoretical" vocabulary.[7]

Feyerabend's main argument against the "theoretical/observational" distinction in vocabulary had been his diagnosis of what happens to candidates for observation terms when theories change: whatever we take those terms' meaning to be, it changes along with theory (cf. Feyerabend 1981 [1957]). If the presumption of stability for the observation vocabulary is rejected, then the question how newly introduced theoretical terms are to be understood, through links to the preceding language used to report on observations, becomes moot. This point, argued with pertinent illustrations from history of science, strengthens the thesis typically associated with Kuhn, that terms introduced originally in the construction of one theory will be differently understood when that theory is displaced in favour of a newer one.

Our entire language, and certainly the part in which observations and measurement outcomes are reported, is through and through theory-laden. So thoroughly infected, in effect, that the idea of a demarcation within vocabulary was just an impossible, even incoherent, fiction.

But this language, our language, we *understand!* So, if theories in science are to be approached realistically, then, to whatever extent they are presented in natural language, they are to be taken literally. For example, if a theory says that there are electrons then that theory is true only if there are electrons (cf. van Fraassen 1980, 11). Scientific language is part of natural language. It can be understood liter-

6 Of these I regard Feyerabend as the most important for the philosophical dialogue. Kuhn discussed a draft of *Structure* with his then colleague Feyerabend in 1961. As I remember Feyerabend, Kuhn, and Hanson, they were philosophical heroes, larger than life. In retrospect, it is clear that this hero-worship marked a division in philosophy at that time, and perhaps this is where the tension between philosophy of science and mainstream analytic philosophy began to form.

7 Suppe was convinced of this (cf. e.g. Suppe 1972), but for different reasons; in later work he criticized Kuhn and Feyerabend severely.

ally, and the efforts to provide it with a verificationist or operationalist interpretation were addressed to an unreal, unrealistic problem.

The main theoretical terms that occur in philosophical controversies about science are not new, not in that sense, not introduced yesterday evening in some radical theoretical departure. They are terms like "electron", "molecule", "gravitational field", "gene", which have long since become part of our natural language in use. You know what it means to say that electrons do not have continuous trajectories in space and time (whether or not you agree that it is so), and so do I. At least we understand that just as well as we understand the assertion that radios receive signals that were emitted far away, or that demons can appear in both male and female form, and we understand it considerably better than we understand universals or propensities.

Both the assertions "Electrons exist" and "Electrons do not exist" are intelligible to us today, even if they would not have been so a couple of centuries ago. Perhaps this is one reason why the very question, whether electrons are real, tends to be met with such a puzzled stare. For our first take on this question could be "Doesn't he know what electrons are?", before we surmise that the questioner knows what electrons are, but wonders whether the theory is true. The important caveat: to understand a term does not presuppose belief that it refers—a very obvious point in ordinary life, but sometimes not obvious in the middle of a philosophical dispute.

Undoubtedly there are exceptions: vocabulary once common but no longer fitting with accepted theory, or on the other hand, terms very recently introduced and still controversial among scientists. But those are not the terms central to philosophical debates about realism, instrumentalism, or empiricism. Nor is the phenomenon of new vocabulary, waiting in the wings for the opportunity to graduate into a proper social role, at all uncommon outside the sciences. It is part of the natural growth of natural language, in which new terms can "become family", so to speak.

2.3 Reaction: David Lewis on Theoretical Terms

Almost immediately after the Received View was meant to have died, in that Illinois Symposium in 1969, it seemed that David Lewis (1970) had successfully resurrected it.

Lewis proposed to save the Received View's view of language by shifting the demarcation from the line between "observational" and "theoretical" to a line between well-understood "old terms" and the "new terms" introduced by the theorists. Thus he posed the problem of interpretation as that of answering the question of what the New terms mean in language that uses only Old terms, the ones that were understood already.[8]

If he could show that the New terms are all explicitly definable, relative to their introducing theory, the entire issue of the theory-ladenness of scientific language would disappear! A main objection levelled to the Received View by those trendy new philosophers of science would be moot.

Lewis followed Carnap and Ramsey in depicting a scientific theory as a finitely axiomatized set of sentences in a language with two distinct vocabularies. The first step toward a clarification, of what the theory says, was then to replace it by its Ramsey sentence. That sentence is formed by replacing all the New terms by variables of appropriate type, and then existentially quantifying those variables. Without loss of generality (let us assume) the New terms will all be predicates, so this is higher order quantification. As example consider a very simple theory with three new theoretical terms:

> Water consists of hydrogen atoms and oxygen atoms.

The Ramsey sentence of this theory is:

> There exist three properties such that water is composed of entities which have the first and third property and entities which have the second and third property.

The justification for regarding this as involving no loss of empirical information is the little meta-theorem that

> all consequences of the original theory which can be stated entirely in the Old vocabulary are also consequences of its Ramsey sentence.

Logically speaking, though, the Ramsey sentence is a great deal weaker than the original theory, and the pattern of properties and relations which it describes can normally be realized in many different ways. Therefore, Lewis proposed to replace the theory not with its Ramsey sentence, but with (in effect) a combination of three postulates: the Ramsey sentence and the Carnap sentence (the postulate that if the Ramsey scheme is instantiated then it is instantiated by the entities named by the relevant theoretical terms), plus a *uniqueness postulate*. The uniqueness postulate says, in effect, that there is a unique selection of properties and relations which realize the pattern described in the Ramsey sentence.

---

8 And he added, quite rightly to my ears: "by 'understand' I mean 'understand'—not 'know how to analyse'" (Lewis 1970, 428).

Lewis's striking conclusion was this:

> Relative to the combination of these three, the Ramsey sentence, the Carnap sentence, and the corresponding uniqueness postulate, the New terms are all explicitly definable using just the Old terms.

Via those explicit definitions, the replacement will logically imply the original theory.

Logically, the combination is stronger than the original theory. One might therefore ask: should we understand the scientist who advances the theory to be asserting something stronger than what he says? To be saying that the Ramsey sentence is actually not multiply instantiated by various sorts of families of properties and relations, but just one? Lewis answers:

> Is there any reason to think that we must settle for multiply realized theories? I know of nothing in the way scientists propose theories which suggests that they do not hope for unique realization. And I know of no good reason why they should not hope for unique realization. Therefore, I contend that we ought to say that the theoretical terms of multiply realized theories are denotationless. (Lewis 1970, 433)

But this attempt by David Lewis to improve on Carnap and Ramsey failed.[9] Its failure follows as a corollary to Putnam's model-theoretic argument, presented in his Presidential Address to the Eastern Division of the American Philosophical Association—the argument that Lewis called Putnam's paradox. This argument showed that any consistent theory with an infinite model will have many non-isomorphic realizations (Putnam 1977, 1980). So, contrary to the uniqueness postulate, there definitely is not a unique selection of properties and relations which realize the pattern described in the Ramsey sentence, not even up to isomorphism.

Lewis clearly came to see this, and he also saw clearly what he had to do, and could do as a realist metaphysician. To avoid this refutation, it suffices to insist that *not just any* sets of entities will be allowed as *properties*, that is, as semantic values for predicates in the language of a theory.

Thus he wrote "New Work for Universals" (Lewis 1983) which offers a metaphysical realist amendment to his attempt. The uniqueness postulated, as newly understood, would be that the Ramsey sentence has a unique realization if the interpretation of the predicates is restricted to *universals* or "*natural classes*" or *nat-*

9 As argued in my "Putnam's Paradox" (van Fraassen 1997b). Putnam's argument applied directly to first-order language, while Ramsey sentences are in second-order language, but the argument applies mutatis mutandis (as is clearly explained in Demopoulos 2003).

ural kinds—these constituting the very special partition of reality that "carves nature at the joints".

Logically, this fixes the argument, but at the cost of introducing notions not found anywhere in the natural sciences. It does not resurrect the Received View but rather replaces it with another view of what scientific theories are, a view that can be articulated only within some version of metaphysical realism.

3 Advent of the Semantic Approach

An entirely new way of engaging with scientific theorizing was needed once the Received View died, and as it turns out, it had been in the making within the sciences itself since the 1930s. But for philosophy there were still obstacles to be overcome.

3.1 Was There Still a Role for a Formal Approach in Philosophy of Science?

The conviction that the language of scientific theories is to be understood literally does not eliminate *interpretation of theories* from the tasks of philosophy of science. In the first half of the twentieth century the interpretation of relativity theory was a live topic: did it require an event ontology, with the concept of an object to be rejected or at best reconstructed in some way? In the second half the interpretation of quantum mechanics came into the limelight. What was to be made of the apparent inconsistency between the theoretical account of a measurement process and the use made of measurement results to link models to observed phenomena? There had been questions of interpretation that Newtonians needed to answer to the Cartesians, and now there were questions that the Bohmians tried to answer to the Copenhagen school.

The most general form for such questions I learned from some lectures in the Pittsburgh Center for Philosophy of Science when I was a student there:

> How could the world possibly be the way the theory says it is?[10]

---

10 Joseph T. Clark, SJ, "Science and Some Other Components of Intellectual Culture", delivered 21 Nov. 1963. (The sentence may not be exactly his, but it comes close enough; cf. Clark 1983 [1966], 304–306). The Center's annual lecture series was predominantly scientific realist and "new wave" in orientation. Besides the very popular lectures by Wesley Salmon, in those years, we also had, for example, Ernan McMullin's "Reflections on Meta-methodology" and Paul Feyera-

Notice that this very question presupposes that the language of the theory is already understood. It certainly does not call for an effort to interpret or re-construe its language so that it no longer says what it seems to say, but rather for a creative effort to reveal ways in which the world of the theory could be envisaged.

To understand a form of discourse, and to understand it literally, does not imply an ability to analyze it. Nor does it imply that the language and its use are utterly transparent to the user. It certainly does not imply that we already have any deep insight into the structure of our discourse, or into its relations to what it is about, its relations to "the world".

At the same time, one reaction one saw sometimes to the writings of Hanson, Feyerabend, and Kuhn was that the formal methods, developed by logicians and exploited by Carnap and his companions, were now to be eschewed in philosophy of science. Countering this, there was another revolutionary development in philosophy at that time, and this provided a new inspiration: *the semantic analysis of logic.* The Dutch logician Evert Beth who had contributed to this approach for both classical and intuitionistic logic, had introduced, in those very terms, *semantic analysis of physical theories.* He had also explicitly linked that to Tarski's theory of truth (Beth 1948, 133).

That is certainly the place to begin. Tarski had begun with a common place in the sciences: *to understand an equation is to know its set of solutions.* A solution to, for example, $3x^2 + 4y = 7$ is the little sequence of numbers $<-1, 1>$ because $3(-1)^2 + 4(1) = 7$. Or as Tarski would put it: this sequence *satisfies* the sentence $3x^2 + 4y = 7$. With that in hand there is then a semantic conception of valid argument:

> The inference from premise $(x + y = 3)$ to conclusion $(y = 3 - x)$ is *valid* because all sequences that satisfy the premise also satisfy the conclusion.

This approach to equations Tarski then generalized to apply to any language at all: the set of sequences that satisfies a complex sentence is a function of the sets of sequences that satisfy its components. Very soon Tarski's semantic analysis had permeated the teaching of classic, extensional logic at all levels, but it also opened the way to a formal analysis of modal and intensional language. This was the insight that Kanger, Hintikka, Kripke, Montague, Scott and their cohort brought to modal logic, and which Beth, just a few years earlier, had proposed as a way to approach physical theory.

---

bend's "The Problem of the Existence of Theoretical Entities" in the spring of 1966. Details can be found at http://www.pitt.edu/~pittcntr/Images/40%20years%20of%20CPS%20Booklet%20SMALL%20OCR2.pdf (last accessed: 30.11.2023).

So here was a task: to investigate the valid inferential relations, semantically defined, in language properly associated with a scientific theory. *These inferential relations may differ from theory to theory* even if the languages overlap. The task set for semantic analysis of physical theory is to display clearly what constitutes the validity of inference in theoretical language, and the tools for it were those created in the semantic analysis of logic, especially of modal logic.

Before this came to be properly explored, however, there was already a tension between two forms of advocacy of formal methods in the philosophy of science, that needed to be resolved.

3.2 Two Analyses of Scientific Theories in the Sixties: Montague and Suppes

There were two impressive, but mutually discordant, formal semantic approaches to the structure of scientific theories already present in philosophy of science in the 1960s. One was Richard Montague's, whose work could surely be seen as a super-Carnapian approach, mechanized and motorized so to speak, with the resources under new development by the logicians' avant-garde. The other was developed by Patrick Suppes with the dissenting slogan that philosophy of science was to use *mathematics, not meta-mathematics.* Both were innovative and promised the use of powerful resources for formal methods. Our hope was to draw on those developments, and yet so as to steer a course between the two extremes of Montague and Suppes.

Determinism and indeterminism had played roles centre-stage in philosophy of physics since the advent of quantum theory and the associated controversies about causality. So Montague's paper "Deterministic Theories" (Montague 1962) should offer a good example of how one could take up a traditional subject with new tools. What does it show about Montague's approach?

That approach was close to the *mathesis universalis* ideal of the Logical Positivists, though linked to the much richer framework of formal semantics. Just as in the Received View, theories are formulated in first-order predicate languages, but rich enough to include a fair amount of mathematics. Montague imposed constraints on the semantics that ensured that the mathematical language had its intended reference. The basic expressions are divided, in the old-fashioned way mirroring the "observational/theoretical" division, into

> *abstract constants* (theoretical constructs) of the language and *elementary constants* (observational terms or sometimes *state-variables*) of the language.

The admissible interpretations are restricted to those in which the mathematical expressions receive their standard interpretation. This was a significant step forward from the orientation toward syntax. Reliance on the distinction between standard and non-standard interpretations of the mathematics is reliance on an understanding outside of the formalism, it is not a distinction that can be assimilated within a formalization.[11]

Montague's very rich language can provide sufficiently complete descriptions of possible trajectories (histories, functions mapping time into states) of a system, and he can define

> A theory *T* is *deterministic* if any two histories that realize *T* and which are identical at a given time are identical at all times. Second, a physical system (or its history) is *deterministic* exactly if its history realizes some deterministic theory. (Montague 1974, 320–326)

Although the languages considered are extensional, the discussion is clearly focused on the possible trajectories (in effect, alternative possible "worlds") that satisfy the theory.

Ernest Nagel and others had emphasized an intimate connection between determinism and periodicity. Intuitively, to say that a system is periodic means that its entire history consists in a series of segments that are exactly like each other. And intuitively, to say that a system is deterministic implies that if it is in the same state at two different times, then what follows those two times (the histories from then on) will also be the same. If that is all there is to determinism, it certainly does not imply periodicity. For then any system that never goes into the same state twice (because of ageing, for example, or of travelling sempiternally, like the Flying Dutchman) would be—vacuously—deterministic and not periodic.

That is pretty well how Montague takes it. A theory that is satisfied only by a single history is deterministic by his definition—vacuously—even if that history is clearly not periodic. So determinism, as he defines it, does not imply periodicity. The point is trivial.

It is still more revealing about Montague's approach to see how he refutes the converse. Regardless of the size of the language, indeed for any infinite cardinality for the language, there is a limited number of axiomatizable sets of sentences (i. e. theories) in that language. Whatever that number is, we can describe a set of periodic histories whose size is larger than that number. Therefore many of these pe-

---

11 This was strongly argued by Beth in his contribution to the Schilpp-Carnap volume; Carnap's response seems to me less than satisfactory. See Beth 1963 and Carnap's remarks in reply to Beth (Schilpp 1963, 871 ff., 926–932). It was also a main theme of Hilary Putnam's presidential address to the Association for Symbolic Logic (Putnam 1980).

riodic histories will not be realizations of any theory in that language, a fortiori, not realizations of any of its deterministic theories! Hence periodicity does not imply determinism, by Montague's definition.

You can imagine how this left us discontented with the orientation toward language. Montague's result is due to defining what the deterministic systems are in terms of *what can be described in a particular language*, while periodicity is defined as a condition on the histories themselves. Does this help us to understand determinism?

Felt discontent with how far the Received View had strayed *mille miles de toute habitation scientifique* will certainly be multiplied by a factor of ten when reading Montague. Virtuoso tour de force, or leger-de-main? We receive a great deal of insight into the power of the new formal methods, but not, it would seem, into the questions about determinism that were so trenchantly involved in the twentieth-century departure from the classical physics worldview.

The same misgivings did not apply to Patrick Suppes' approach. Suppes followed a procedure typically found in contemporary mathematics: the axiomatic method is revised, or replaced if you wish, by defining the kind of system the axioms meant to characterize. So for example, the traditional axiomatic development of Euclidean geometry is replaced by the definition of Euclidean spaces as a class of mathematical structures. Euclid's theorems are then proved as statements which are satisfied by all those structures. If Euclidean geometry is adequately axiomatized as a theory, then those structures are precisely the models of the theory. But to study the class of Euclidean spaces, one need not assume that the theory can be axiomatized—just as we can study the natural numbers without assuming, contrary to fact, that arithmetic has a consistent and complete axiomatization.

In a similar manner, Suppes replaces Newton's laws, which had been explicated as axioms in a language in various ways, by the defining conditions for a set of structures to count as Newtonian systems of particle mechanics. To study Newton's theory is, in Suppes' view, to study this set of mathematical structures. Suppes and his co-workers provided further guiding examples through his work on the foundations of psychology (specifically, learning theory), as well as of physics (specifically, classical and relativistic particle mechanics).

In addition, Suppes was equally intent on re-focusing the relationship between pure mathematics and empirical science, starting with an article that argued that the conception of model is essentially the same in both (Suppes 1960, 1962). While still in a "formal" mode, at least as compared to the writings of a Maxwell or a Sellars or Kuhn, Patrick Suppes had clearly moved away from preoccupation with formalism and logic, to be much closer to the actual scientific practice of the time.

Nevertheless, there were misgivings with this approach as well. It seemed that by forsaking any attention to the language of science, this approach would disengage from a number of central problems in contemporary philosophy of science. For example, the debate concerning conventionality in the foundations of relativity theory, from a formal point of view, hinged on questions of definability and axiomatizability—meta-mathematical questions. The problem of counterfactuals, and such concepts as laws of nature, physical necessity and other modalities, were at least intimately connected to issues concerning the language of science, a subject not broached by Suppes's methods.

At the same time, it will be clear that our sympathy was on the side of Suppes rather than Montague. Suppes' slogan "mathematics, not meta-mathematics!" had a strong appeal, even if it had to be kept within limits. Nevertheless, the question was now how to set a course between these two extremes. I ended this discussion of Montague and Suppes with:

> To avoid both Scylla and Charybdis, I shall now outline an approach to the formal structure of theories that does not place much emphasis on syntax or proof theory, but yet makes it possible to focus on certain linguistic structures associated with given theories. (van Fraassen 1972, 311)[12]

3.3 Frederick Suppe: Semantic Approaches Alternative to the Received View

The long introduction of Suppe (1974) is first of all a detailed history of the development and problems of the Received View from the 1920s till the symposium he organized in 1969. At the end Suppe examines alternatives to the Received View that had come on stage, in the section "Semantic Approaches" (Suppe 1974, 221–230).

To begin, Suppe insists that a theory is not a collection of statements or propositions, but an extra-linguistic entity that admits a number of different linguistic formulations. But that is sufficient to suggest the usefulness of model-theoretic concepts and techniques. Moreover, the presentation of a theory often takes the form of specifying an intended model of its equations. As he states the main themes of the new outlook, they were (in only slightly amended phrasing):

> Theories are extralinguistic entities which can be described by their linguistic formulations.

---

12 This paper was presented in the University of Pittsburgh Center for Philosophy of Science Annual Lecture Series in January 1970.

> The semantic techniques of model theory in logic will be useful in analyzing the structure of scientific theories.
>
> A scientific theory has as its subject matter a class of phenomena, its intended scope or domain; it will abstract certain parameters from the phenomena and characterize them in terms of just those parameters.

It would be more apt to say "represent" rather than "characterize", since an assignment of values to a certain set of parameters is an abstract structure, it is not a phenomenon but a selective representation of a phenomenon. Suppe goes on to discuss those representations, using the term "physical system" rather than "model", since he has a specific sort of theories in mind.

> [The] important function of experimental design is to determine the representation of phenomena by physical systems. (Suppe 1974, 226)

We can read this as meaning "by data models", since that passage refers to Suppes 1962. However, there was a parting of the ways here, between Suppe and myself. Although in this section Suppe describes his and my own forms of semantic analysis as akin—as indeed they were—a resolute scientific realism and modal realism inform his presentation. For him the proximate subject for theorizing is not the phenomena themselves, but physical systems characterized by those parameters alone. These systems are idealized replicas, and Suppe describes their relation to the phenomena in terms of a counterfactual, a "would have been". The emphasis on the counterfactual connection between the real empirical phenomena and the models of the theory becomes a main theme in his later book (1989), to mark his specific variety of scientific realism. But that crucial reliance on objective modality was already clear in an earlier paper:

> On this account a theory may be construed as defining a class of theoretically possible physical systems; the theory will be empirically true just in case this class is identical with the class of causally possible physical systems. (Suppe 1971, 14, fn. 15)

Other practitioners of the semantic approach in the coming decade also took scientific realist positions, and it was surely a value of this approach that it lent itself equally well to realist and empiricist views of science.

Although the *identification* of a theory with a class of systems can easily be criticized, that a theory is presented by presenting such a class of structures as candidates for representation of the phenomena in the theory's domain, was common ground for us.

Nevertheless, there was another disparity, which may seem trivial at first blush, but probably helped to obscure some of the debates over realism. Suppe and I had a slightly different way of thinking of the class of models presented when a theory is presented. For Suppe, this was precisely the class of physical systems, characterized by all and only the quantities that appear in the theory's equations —the *relevant* quantities—and satisfying those equations. These are certainly the systems that the scientific theorizing would be all about, I grant that. But in my way of thinking about models, extensions of models of a theory are also models of that theory.

As illustration, think of a game of billiards, and imagine a set of equations that precisely describe the mechanical behaviour of the billiard balls, in a certain interval of time. The terms in those equations refer solely to the mechanical quantities: mass, force, velocity, ... And there exists, mathematically speaking, a physical system (using Suppe's term here) which satisfies those equations. Other characteristics of the billiard balls, such as colour, ownership, country of origin, and what have you, do not appear. But the model (that physical system) has many extensions, in which those *irrelevant* quantities can also be assigned values. Those extensions, infinitely many of course, are in my way of thinking and speaking also models of the theory, and models of that piece of history of the billiard balls. All the phenomena, relevant and irrelevant, occurring in that piece of history of the balls, will therefore be accurately represented in some model, some extension of the physical system in question, of the theory.

This is perhaps more typically a logician's way of thinking; it may just show "where I am coming from", as they say. But there are times when it makes a pertinent difference.

3.4 Language: How It Is Relevant and Irrelevant to Theory

When the semantic approach is criticized, it is often because of its perceived extreme dismissal of language as a topic for philosophy of science. That is a caricature.[13]

What was rejected in the Received View was *not* that there are significant relations between models of physical systems and language that describes those systems. What was rejected was the idea that it was fruitful to study the formal struc-

13 The following passage from *Laws and Symmetry* was variously quoted (e.g. in Halvorson 2012): "if the theory as such, is to be identified with anything at all—if theories are to be reified—then a theory should be identified with its class of models" (van Fraassen 1989, 222). But that sentence (note its conditional form!) was part of a brief exposition of Patrick Suppes' approach in the fifties, a precursor to the semantic view, and taken out of context if used as a purported statement of the semantic view (cf. Suppe 2000, 111).

ture of a sort of *language in which the theory as a whole can be formulated.* That would mean a language in which, in some form, in effect, everything can be expressed that this theory implies about the phenomena, the models, and the relation between them.

Such was indeed the dream format of a scientific theory developed as an instance of a Hilbert-style axiomatization or of a *mathesis universalis* such as *Principia Mathematica* purported to be. As mentioned, Stephen Toulmin's historical account of the origins of the Received View lists this dream as the overarching ambition of the Logical Positivists. I quoted part of it above; Toulmin added:

> On this program, the value of the axiomatic model was not just heuristic or epistemological: it now became the obligatory form for expounding a coherent and unified scientific theory, or *Weltbild.* (Toulmin 1974, 602)

Is it also necessarily a form in which any scientific theory can be presented, in principle? Leaving aside the limitations of the sort of language they must have had in mind, I suppose so. Since scientists create and communicate their theories, all the required resources must be there in some part of natural language, supplemented by elements that can be introduced within natural language. At least to communicate: not to characterize or define adequately and completely, on pain of contradiction. Well, fine. Whatever we must grant here certainly does not imply that to understand the sciences it is crucial to require a formal reconstruction of those natural language resources, or even to be thinking about it, for that matter.

The scientist offers his equations, whose terms stand for physical quantities, to show relations between the values of those quantities. Equations are sentences, so there is an important fragment of our language that is explicitly used by the scientist in the presentation of models to represent the phenomena. Why shouldn't our focus remain on that fragment, and more specifically on how it relates to the models thus presented, rather than on the imaginary formalization, in some logically regimented language, of the whole theory?

That idea was hard to appreciate at the time. A good example of how a preconception may influence a reading is Alonzo Church's review of Birkhoff and von Neumann (1936) for the *Journal of Symbolic Logic.* Church placed their construction of the logic of quantum mechanics in the truly inappropriate context of the problem of formalizing a language adequate to the entirety of science:

> It has been maintained by various writers that the selection of one system of logic as against another with laws in part different is determined, not by a priori considerations, but partly by necessities of the intended application and partly by considerations of convenience and elegance […]. It is, of course, a corollary of this position that the advance of an experimental

> science may some day require revision of the system of logic on which the mathematical theory of the science is based.
>
> Apparently taking this for granted, the present authors suggest that this situation has already arisen, in that quantum mechanics is best formulated in terms of a logic in which the distributive law [...] and its dual, fail.
>
> The authors succeed in making this suggestion plausible. There remains, however, grave question whether a logic of this character can be made in any reasonable degree adequate, and until (and unless) this question is settled affirmatively their suggestion cannot be regarded as indicating more than a vague possibility. An answer is clearly not to be expected within the bounds of a single paper of this length. But it is to be regretted that they have confined their attention entirely to what they call the calculus of propositions (really an analogue of the Boolean algebra of classes), and give no indication how quantifiers might be introduced so as to obtain adequate use of them without re-establishing the distributive law. (Church 1937, 44–45)

That this "regret" was quite beside the point seems not to have been remarked in the philosophical literature till much later. The only clear, early reaction came from Beth:

> [I]n my opinion the so-called *'problem of developing quantification theory within the logic of complementarity'* is less urgent than some logicians seem to believe. Curiously enough, as far as I know this problem has been raised exclusively by reviewers in the *Journal of Symbolic Logic*. I wish to emphasize that, for the purpose which in classical logic is served by quantification theory, quantum theory applies the procedures known as composition and second quantization. (Beth 1960, 74)

It was precisely the sort of analysis of physical theory by Birkhoff and von Neumann, that Church misunderstood as a sketch for a *mathesis universalis*, which set the program for the semantic approach to science.

3.5 The Origin of the Approach in von Neumann, Birkhoff, and Weyl

The story for both Fred Suppe and me began with Johann von Neumann, even if not quite in the same way.[14] Already in his book on the mathematical foundations

14 Suppe was a student of Arthur Burks, who introduced him to much of von Neumann's work, also outside of the subject of foundations of quantum mechanics, and Suppe saw von Neumann as offering a paradigm example: von Neumann's proof that wave mechanics and matrix mechanics are equivalent formulations of quantum theory essentially involves this approach, though he never attempted to extend his techniques to any general semantic analysis of theories (see Suppe 1974, 222).

of quantum mechanics (von Neumann 1955 [1932]) and then in the famous joint paper with Birkhoff in 1936, von Neumann had begun what we can see today as a semantic analysis.

This analysis pertained to a restricted sort of language closely identified with a scientific theory. The statements in such a language can all be expressed in the same form: "quantity m has value r". When a physical system is represented within the framework of a given theory, such a statement is true in some possible states of that system and false in other states. The proposition expressed by such a statement is then represented (or coded, if you like) by the mathematical representation of the states in which it is true.

Looking into those representations, Birkhoff and von Neumann showed first of all how we can view some such statements as functional composites of others, in the way logicians find familiar. The family of propositions expressed then has the structure found in classical logic in the case of classical physics, but a quite different structure in quantum physics.

It was Hermann Weyl, in a not much noticed paper, "The Ghost of Modality" (Weyl 1940) in a festschrift for Edmund Husserl, who pointed out that this analysis parallels ways in which the semantic analysis of classical, modal, probability, and intuitionistic logic can be presented. The general insight we find early on:

> We are here concerned with propositional functions $A(x)$ or predicates referring to an arbitrary element x in a certain field or space ω of individuals or points. For instance, x may range over all integers or over all points of a geometric space. If we choose as our space the phase space of a physical system, then the argument x indicates any phase or state of that system. Complete knowledge of a point x consists in knowing its position; any sort of incomplete knowledge, in knowing that it lies in a certain region α of the space ω. (Weyl 1940, 281; one font for a mathematical symbol adjusted)

Then he explains argument *validity* in terms of inclusion between the corresponding regions. My use of the term "semantic analysis" is anachronistic here; but I insist that the point is correct. Weyl does not succeed in his attempt to display, in completely unified fashion, the analysis of statements in terms of their relations to sets of states for each of those logics. He is not very sympathetic to C. I. Lewis' modal logic. But he knows how Gödel related that to Heyting's intuitionistic logic, and how the latter can be analyzed along the above lines using a topological space. The general scheme is shown at the end to be precisely that introduced by Birkhoff and von Neumann.

What is crystal clear in his (even if, to our ears, somewhat dated sounding) paper is that the semantic analysis of statements, significant for a theory, consists in a linking of such statements A to mathematical objects $|A|$ representing their semantic value. Then the implication relationship (argument validity) gets its obvious representation as inclusion:[15]

A1, ..., An *imply* B if and only if the intersection of |A1|, ..., |An| is included in |B|.

In the more familiar cases that have nothing to do with physics specifically, we are used to thinking of statements as sentences composed of other sentences, with the composition paralleled in the semantic values, exhibited in equations of the form

$$|f(A1, \ldots, An)| = f^*(|A1|, \ldots, |An|)$$

where the functions f* are definable within the mathematical modelling framework of the theory. The further desideratum is that it be reasonable to link the connectors f with some familiar connectives in classical logic; for example

$$|A \,\&\, B| = |A| \cap |B|.$$

For a philosophy student in the 1960s this was precisely the familiar pattern of the new and exciting semantic analysis of modal logic. There the semantic value |A| would be a set of "possible worlds", with equations defining a connective like this:

$$|\Box A| = \{w\text{: all } w' \text{ such that } w \text{ bears } R \text{ to } w' \text{ belong to } |A|\}$$

where the right-hand part could then be written in functional form as □*(|A|). That is, the connective □ has its own semantic value, a function □* defined on sets of possible worlds.

But what is distinctive of the Birkhoff and von Neumann approach, as is clear also in Weyl's paper, is that the typical logician's preoccupation with *syntactic* form is there simply left aside. All the significant statements can be expressed in the same form:

(m, r), e.g. the mass of the system, in kg, equals 15

or

(m, E), e.g. the mass of the system, in kg, is in interval (13, 17)

15 For simplicity I take the semantic values to be sets; a more general formulation would take them to be elements of lattices or algebras of any sort.

Then the functions that relate such statements are, so to speak, inside the formalism:

> the conjunction of (m, E) and (m, F) is (m, E ∩ F)
> e.g. the conjunction of (m, [0,1]) and (m, [1,2]) is (m, [1])

With this simplicity of syntactic form, the emphasis can then be entirely on the side of the semantic values. Equations can also be regarded as such statements; for example (f = ma) can be understood as (f – ma, 0).

Was any of this appreciated by philosophers of the 1930s? Note the glaring difference between this approach to language of theories and the formalism of the Received View. It is not just that the overly syntactic form was annoying to later eyes, but that we encounter here a quite different identification of language worth analyzing.

To sum up: for von Neumann, Birkhoff, and Weyl the statements to focus on all have the same basic form: *quantity m has value r*, or more generally, *quantity m has value in Borel set E*. We called these the *elementary statements* (cf. Suppe 1974, 228–229). Some of these are indeed conjunctions or negations or disjunctions of others, not because of their form, but because of relations among their semantic values.

A language restricted to these statements is *not* a language designed to formulate the whole theory, in the way of the Received View of Theories. Rather, the associated language is a tool to describe the systems that the theory is about. Its semantics links the statements deemed significant to the models that the theory offers for representation of those physical systems.

3.6 Beth: Semantic Analysis of Physical Theories

Above I asked: Was this appreciated by philosophers of the time? The answer has to be No, with just that one somewhat belated exception already mentioned: the Dutch philosopher-logician Evert W. Beth whose paper "Semantics of Physical Theories" I quoted above. During the war years he had been writing a little textbook, *Natuurphilosophie*, which to begin he certainly conceived as basically along the same lines as any such book would be written by a Logical Positivist. But just prior to publication (after the war, when paper was available) he added a note to his chapter on the structure of quantum mechanics:

> On rereading the material, which was written several years ago, I lean to the opinion that it is possible to give a semantic construction to the logic of quantum mechanics with the help of

> Hilbert space [...]. The interpretation [that he had given, in terms of experimental results] is *not* analogous to semantics. (Beth 1948, 133)

Then he developed the approach of Birkhoff and von Neumann briefly in three articles (Beth 1948/1949, 1949, 1960), with specific examples of how to analyze implication among statements that describe physical systems. I imagine that his brevity can be accounted for by his taking the earlier literature to be entirely clear and transparent to the reader. But in this he overestimated the philosophical public.

Beth's thick yellow book on the foundations of mathematics and logic had been a sort of bible for me as an undergraduate, and bit by bit I acquired other books and papers by him—he was one of my heroes. I began a more general development of Beth's ideas in "On the Extension of Beth's Semantics of Physical Theories" (1970). The framework I presented there I then used to provide a semantic analysis of quantum logic along the lines of the semantic analysis of modal logic.

The generality that I proposed was to think in terms of a special sort of languages, that I called *semi-interpreted.*

The idea was this. When we look at the language of "elementary" statements, that is, the statements that attribute values to physical quantities, their semantics should be seen to have two faces. On the one hand, they receive a complete interpretation in the models of the given theory. On the other hand, they may receive indirectly a connection to the physical phenomena, because the theory will locate those phenomena in its models.

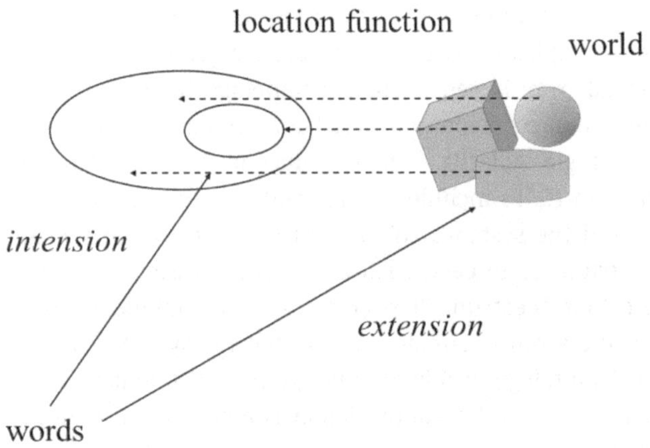

Fig. 1: Semantic relationships between language, models and the world.

What quantum logicians call the *logic* of that language derives entirely from the interpretation in the space of models, which represents the possibilities admitted by the theory for systems in its domain. But what makes it the language of a *physical* theory is the relation to embeddings of the phenomena in the models (see Fig. 1):

- To understand the inferential relations in scientific language we need to see how the physical quantities under discussion are represented in the models provided by the relevant theory.
- In traditional "modern" logic, inferential relations were seen as derivative from the *intension* of the terms; here the intension of a term is its semantic value, which is defined on the space of models of the theory.
- Scientific models are offered as representations of physical systems, representing them *as* having the structure displayed in those models, thus yielding indirectly the *extension* of the terms.

Note that a semantic analysis is not an analysis of how meaning is bestowed: it is an analysis of an already fully meaningful language. Unlike in the Received View, we are not labouring under the image of language that is in itself unintelligible (or at least, devoid of meaning), which needs to receive meaning through correspondence rules.

As an example, let us take discourse of a sort we all understand very well, language in which we attribute colour, with such statements as 'The apple is red', 'The road is yellow', 'The entire city is emerald green'. We know how to form such statements and know the roles they play in inference. For example, if the city is green then it is not red. A semantic analysis of this well-understood discourse, along the above lines, begins with the three-dimensional colour spectrum (hue, saturation, intensity) and treats those examples I just gave as locating objects in that colour space. That nothing green all over is red is then a tautology in this language, since the regions in the space that correspond to "red" and "green" do not overlap.

Put more technically, a single quantity, C, takes values in that spectrum, and elementary statements of form (C, E) include e.g. (C, red): the system's colour is red. This is true if and only if the system's colour location is in the area E (say, the "red" area) of the spectrum. So to each such statement is assigned a region of that logical space, the colour spectrum. It is clear that the conjunction of (C, red) and (C, green) is itself the absurd statement (C, $\emptyset$) that the system is located in the empty set of points. A tautology in this little language is a statement whose semantic value is the entire space, and a contradiction is a statement whose semantic value is the empty set.

To the exposition of Beth, in which I drew on this conception of semi-interpreted language, I added a general form for dynamic and statistical laws imposed on such a state-space, as well as some discussion of physical modality.

3.7 Departure from Suppes: The State-Space Approach

Patrick Suppes' set-theoretic approach was intended to be of such generality that it could cover group theory and Euclidean geometry as well as Newtonian or relativistic mechanics. Although the use of formal methods in Patrick Suppes' practice was clearly close to the semantic approach, it was certainly also quite different from what we had seen in the Birkhoff–von Neumann–Weyl–Beth view of quantum mechanics. Eventually I commented briefly on the disparity:

> While I consider Suppes's account of the structure of scientific theories an excellent vehicle for the elucidation of these general distinctions, I do regard it still as relatively shallow. In this I am mainly concerned with the relation between physical theories and the world rather than with that other main topic, the structure of physical theory. With respect to the latter I see two main lines of approach: one deriving from Tarski and brought to maturity by Suppes and his collaborators (the *set-theoretic structure approach*) and the other initiated by Weyl and developed by Evert Beth (the *state-space approach*). The first is adamantly extensionalist, the second gives a central role to modality. Each was somewhat language-oriented to begin with, but both shed these linguistic trappings as they were developed. (van Fraassen 1980, 66–67)

Really, that was a bit too strong; there was a mutual accommodation possible. One way was to think of Suppes' presentation of classical particle mechanics in this way:

> The set of models Suppes defines is itself the logical space, and the individual models are the points in that space.

Then the statement that e.g. the mass of the system is in interval [a, b] is assigned as its intension the set of models in which the total mass is in that interval.

Another way, which seemed more realistic to me, is to think of Suppes as implicitly describing the general structure of phase-space, and that for each dimensionality, there is a logical space in which to locate particle mechanics systems, namely the phase-space with that dimension. That would then be analogous to the description of the general structure of Hilbert space with Hermitean operators for systems of quantum mechanics.

These examples suggested a general format, though we do need to make a division between relativistic and non-relativistic theories. In the latter, the systems are physical entities developing in time. They have accordingly a space of possible states, which they take on and change during this development. Think for example of a simple harmonic oscillator: a classical particle moving along a straight line subject to a force proportional to the distance from, and directed to, a fixed point 0 on that line. Using q and p for its position on the line and its momentum along that line, its motion is described by equations of the form

$$dq/dt = (1/m)p;\ dp/dt = (-1/n)q.$$

At each time t, the particle has a state represented by a point in the q-p plane (its state-space or phase-space). In time, the particle describes an "orbit" or "trajectory" in this plane, described by the equation

$$E = \frac{1}{2n}q^2 + \frac{1}{2m}p^2 = \text{constant},$$

E being the mechanical energy. The physically possible trajectories are the curves which satisfy this equation.

Noting the parameters with unspecified values in these equations, this introduces the idea of a cluster of models united by a common state-space. Each such model has in addition a domain of objects plus a "history function" which assigns to each object a history, i. e., a trajectory in that space. A real theory will have many such clusters of models, each with its state-space. So the presentation of this sort of theory must proceed by describing a class of state-space types.[16]

3.8 Connection to the Quantum-Logical Approach to Foundations of Physics

At that time, much of our attention in philosophy of physics was directed to the then newly invigorated "quantum-logical" foundations of quantum mechanics, which was being developed by physicists and mathematicians. An exhilarating series of conferences in which philosophers also took part, were taking place at the University of Western Ontario, as well as in Italy, Finland, and Germany.

In the simple example I gave of a colour space there is no problem identifying the semantic values of conjunction, disjunction, negation: the logic is classical. In the case of the logical space presented in quantum mechanics, following Birkhoff and von Neumann, however, the propositions (semantic values of the elementary statements) do not form a Boolean algebra.

---

16 In the case of relativistic theories, early formulations can be described roughly as relativistically invariant descriptions of objects developing in time—say in their proper time, or in the universal time of a special cosmological model (e.g., Robertson-Noonan models). A more general approach to relativistic theories, that was explored by Clark Glymour and Michael Friedman, takes space-times themselves as the systems.

The standard way in which the elementary statements in quantum mechanics were understood was introduced explicitly in von Neumann's book (von Neumann 1955 [1932]) and exploited in Birkhoff and von Neumann (1936). Looking back with hindsight we would now say that it incorporated the *"eigenstate-eigenvalue"* principle that later came to be questioned (as we will see below). That is the principle that the proposition that physical quantity m, of a given system, has value r precisely if that system is in an eigenstate of the representing operator M for eigenvalue r. Following the usual form of semantics, the statement that m has value r, written as (m,r), has as semantic value a proposition, identified as the set of conditions in which it is true:

$$\begin{aligned} |(m, r)| &= \text{the set of eigenstates of m corresponding to eigenvalue r of m} \\ &= \{\psi: M\psi = r\psi\} \end{aligned}$$

This can be given a more general form to accommodate intervals of values by use of projection operators M_E taken from the spectral decomposition of the operator M:

$$|(m, E)| = \{\psi: M_E\psi = \psi\}$$

which in quantum mechanics, via the Born rule, is the set of states in which a measurement of m will yield a value in E with certainty.

In a given (Hilbert) state-space for a system these propositions are precisely the subspaces. Operations meet and join can be introduced, e.g.:

$$\begin{aligned} S \wedge T &= \text{the largest subspace contained in both S and T} \\ &= \text{the intersection of S and T} \\ S \vee T &= \text{the least subspace that contains both S and T} \end{aligned}$$

That result of the join-operation, however, is larger than the set-theoretic union. The lattice of subspaces of a Hilbert space is non-Boolean. Various ways of exploring the resulting inferential relations among the elementary statements became the subject of the "quantum-logical" approach to the foundations of quantum mechanics.

3.9 The Quantum Mechanics Interpretation Wars

Interpretation of quantum mechanics was beset by polemics from the very beginning, and throughout.[17] On the one side there were, for example, Leon Rosenfeld's life-long insistence that there could be no feasible alternative to the way the theory had been understood in Copenhagen at the outset. On the other side were such verdicts as John Bell's on von Neumann's argument against hidden variables: "not just flawed, it's silly!", "not merely false but foolish".[18]

Too much gets lost in polemics. Since the semantic approach lent itself to a more technical and less partisan treatment of interpretation, R. I. G. Hughes (a student of Jeffrey Bub) and I were intent on making both hidden variable and no-hidden variable alternatives rigorous and precise.[19]

3.9.1 First Controversy: The Line Between Nature and Observer

First of all we needed to confront the most familiar objection, that the Copenhagen point of view, at least as presented by Nils Bohr, made the notion of measurement primitive, as it were, outside of the domain of quantum physics. Bohr had insisted on a division in nature between the realm modelled in the quantum theory and the "classical" macroscopic world in which scientists do their work. Whatever Bohr meant—his writing is an easy target!—that would be a debilitating feature of the theory.

How did that objection retain its appeal for so long? Von Neumann had given the correct response already in the 1930s.[20] *There is no dividing line in nature,* but only one drawn by the modeller, who chooses what to model.

When a physicist models a part of nature, s/he can draw the line wherever s/he likes. An apparatus, a spectrometer, a Geiger counter, an electron microscope, a cloud chamber, or a box containing a cat and lethal device, *can be either included*

---

17 See the recent popular science book *What Is Real?* by Adam Becker (2018), which is a good guide especially to all the invective ever written about the "Copenhagen interpretation".

18 I had learned a great deal about this early on, from my colleague Henry Margenau at Yale, and from being on the committee of Margenau's student D. L. Reisler's dissertation on the Einstein-Podolsky-Rosen paradox at Yale in 1967.

19 Hughes eventually favoured a specific form of interpretation, the quantum event interpretation (Hughes 1989, 301–306). That was closely related, in different ways, to Jeffrey Bub's quantum-logical approach and to my "Copenhagen" option (Hughes 1989, 306–308, 315–318).

20 "Moreover, it is rather arbitrary whether or not one includes the observer in **M**, and replaces the relation between the **S** state and pointer positions in **M** by the relations of this state and the chemical changes in the observer's eye ..." (von Neumann 1955 [1932], 352).

or excluded, wholly or in part, from what is to be modelled. But however you draw the line, what you are modelling quantum-mechanically is what is on the other side of that line.

This was von Neumann's response which, in effect, defined the "external observation" interpretation, to use the name John Wheeler gave it. Despite more radical sounding pronouncements, including some due to Bohr himself, this was the upshot of what was generally accepted as the "Copenhagen interpretation", in the physics community.

John Wheeler himself greatly admired Bohr, and gave this beautiful, concise, sympathetic presentation of the "external observation" view:

> The external observation formulation of quantum mechanics has the great merit that it is dualistic. It associates a state function with the system under study—as for example a particle—but not with the *ultimate* observing equipment. The system under study can be enlarged to include the original object as a subsystem and also a piece of observing equipment—such as a Geiger counter—as another subsystem. At the same time the number of variables in the state function has to be enlarged accordingly. However, the *ultimate* observing equipment still lies outside the system that is treated by a wave equation. As Bohr so clearly emphasizes, we always interpret the wave amplitude by way of observations of a classical character made from outside the quantum system. The conventional formalism admits no other way of interpreting the wave amplitude; it is logically self-consistent; and it rightly rules out any classical description of the internal dynamics of the system. With the help of the principle of complementarity the external observation formulation nevertheless keeps all it consistently can of classical concepts. (Wheeler 1957, 463)

But it was Wheeler himself, in this 1957 article (in which he presented Everett's "relative state" formulation) who sounded the clarion call for a major shift in view.[21]

3.9.2 Paradigm Shift: The World as a Quantum-Mechanical System

The "external observation" interpretation implies an obvious limitation for the theory: on this view, to treat a system as closed and isolated can never be more than an idealization or approximation, or useful fiction. Specifically, as Wheeler argued, the "external observation" interpretation does not allow for cosmology

21 Russian physicists, such as Blokhintsev, had rejected the Copenhagen view from the beginning, as putatively idealist, irreconcilable with dialectical materialism. They insisted that quantum mechanics was a true *mechanics*, modelling mechanisms in nature with no special exceptions for observation or measurement. Surprisingly their view was taken up wholeheartedly in foundations of physics, in the West, in the midst of the cold war.

as part of physics (Wheeler 1957, 463). For there the models are of the entire world, thus including apparatus and physicist and everything else, with no "this side of the line" remaining.[22]

That may not be a worry for practice in either laboratory or industrial application. But it certainly raises a conceptual puzzle. The infamous *measurement problem* is the problem of reconciling two representations that could both be apt:

(A) a model of a system X, as a closed system, on which measurements by apparatus Y have had specific outcomes, which are eigenvalues of the measured observable

and

(B) a model of that very process in which X was subject to measurement by apparatus Y, with X+Y treated as a closed system

Reconciling (A) and (B) appeared to be impossible: the outcomes mentioned in the former do not appear to have any place in the evolution of the quantum-mechanical state of a closed system. Von Neumann understood this, and offered his second response: his Projection Postulate, the postulate that in a measurement the state of the measured system "collapses".[23] The effect of the measurement on the measured system is to change its state into an eigenstate of the measured observable, and the corresponding eigenvalue is what the outcome registers.

As a tool for calculations in the application of quantum mechanics, this Projection Postulate works perfectly well. And in the external observation view, the problem of reconciling (A) and (B) could be set aside, since modelling a closed system was anyway, properly understood, an oxymoron.

Henry Margenau rejected this solution already in his response to the Einstein-Podolsky-Rosen paradox in 1936. But his was a voice in the wilderness. After 1957 however, practically all discussion of foundations of quantum mechanics shifted in the way Wheeler advocated. Not just Everett's relative-state formulation, but all the main new interpretations to come on the scene (Bohm, GRW, coherent histories, etc.) took the same line as Wheeler. Now the measurement problem could not be set aside: measurements happen inside closed or isolated systems, specifi-

22 This conception of cosmology itself is now in question; it is challenged by Carlo Rovelli: "physical cosmology is not the science of everything: it is the science of the largest scale degrees of freedom" (Rovelli 2017, 10, fn. 3).

23 Von Neumann does not use the terms "collapse" or "Projection Postulate" when he introduces this, but writes "On the other hand, the state [...] undergoes in measurement a non-causal change" (von Neumann 1955 [1932], 418).

cally, if the world as a whole is such a system. So Wheeler saw as a major advantage for Everett's interpretation, that it has a place for observations only insofar as they take place within the isolated system. The theory of observation becomes a special case of the theory of correlations between subsystems (Wheeler 1957, 463). That was, more or less, the way in which each new interpretation had to finesse the problem.

3.9.3 Honouring, Nevertheless, The Copenhagen Legacy: Via a Modal Interpretation

At that time I too accepted Wheeler's shift in view as the norm: we must depict the theory as in principle providing models for the universe as whole, with measurement processes all included.[24] For as mentioned above, I took the question to be addressed in any such attempt to be

> how could the world possibly be how the theory says it is?

Nevertheless, I set myself the goal of constructing an interpretation that would preserve what I saw as true insights among the Copenhagen ideas about how the world could possibly be as quantum mechanics models it. The first of these was:

> **I. Indeterminism:** The world modelled in quantum mechanics is genuinely indeterministic, with a quantum-mechanical state and its evolution constraining what is possible or probable but not, beyond that, what is actual.

The second idea that I saw as also central to the Copenhagen ideas was that of complementarity, in its minimal form: certain pairs of physical quantities literally cannot have simultaneous precise values. In view of how this was related to the interpretation of states and measurement, that amounted to:

> **II. Complementarity:** Nothing can be true together that cannot have probability 1 together.

These are intuitively phrased principles, which could be made specific in different ways. I took it that physical *quantities* will have or take actual values, which are displayed in measurement outcomes. Furthermore, looking at the Born rule that calculates the probabilities of measurement outcomes, the role of the quantum

---

24 I did not see a feasible alternative until the appearance of Carlo Rovelli's interpretation, his "relational quantum mechanics" (cf. Rovelli 1996; van Fraassen 2010).

state must be to constrain what is possible or probable. And from that rule I also took it that the only values that a quantity can take are the eigenvalues of the representing operator.

Thus read, these two principles together imply that two quantities can have simultaneous values only if their representing operators have eigenstates in common. When that is not the case, then at any time either one has a value and the other does not, or neither has a value (*indeterminacy*).

Does a quantity ever have a value not displayed in a measurement outcome? That is certainly left open by the above two principles. Specifically, it is consistent to add that if the quantum state of the system is an eigenstate of a quantity, then that quantity has the corresponding eigenvalue, whether or not any measurement is performed. In words echoing Einstein,

> **III. Certainty:** If a measurement of quantity A would yield value *a* with certainty then A actually has value *a*.

This is just one half of the "eigenstate-eigenvalue" link that I described above in the analysis of quantum logic. But the other half of the "eigenstate-eigenvalue link" cannot be consistently added. The state of a closed isolated system evolves deterministically, in accordance with Schrödinger's Equation. If the truth values of the elementary statements were uniquely determined by the state, and if these statements described all and only what is actual, it would follow that all that is actual is also uniquely determined. That would run into trouble with principle I: *indeterminism.* Hence, if the world as a whole can be modelled as a quantum-mechanical system, *we cannot add the converse* of III.

The only way to go on then is to *separate quantities from states,* that is, to allow a physical quantity to *actually* take values even when the state is not an eigenstate of the operator that represents it.[25] That does not eliminate indeterminacy, for II remains intact.

Specifically, to address the measurement problem, we can assert that at the end of a measurement, the "pointer observable" takes an actual value: the outcome

25 States and quantities are related of course, but states cannot be identified as quantities in this context. A system is in a pure state if and only if it is in an eigenstate of a certain quantity, so technically, dealing with pure states is no different from dealing with projection operators. But the same cannot be said of mixed states. An important associated theme was that mixed states cannot in general be thought of as representing ignorance of pure states (cf. Cartwright 1971).

of the measurement. This is in line with the principles displayed in every quantum mechanics textbook.[26]

So we must now distinguish two sorts of elementary statements to be associated with quantum theory. We can keep the notation (m,r) that I had above, with its meaning—call that sort of statement a *state-attribution.* We must then not read it as saying that m has value r, but rather that the state is such as to guarantee that (*make it necessary that*) m has value r. When (m,r) is true, which means that the state is an r-eigenstate of m, then a measurement of m will or would yield value r with certainty, and principle III applies.

But when (m,s) is not true for any value s, it must be allowed that m might nevertheless *actually* have value r. I gave that statement the notation <m, r> and call it a *value attribution.* Thus <m,r> is true exactly if m actually has value r. Principle III then says that (m,r) implies <m, r>; but the converse is not asserted.

I called this a *modal* interpretation, since it views the quantum-mechanical state and its evolution as merely constraining what happens, with many possibilities left open for the values that quantities may have—including of course the values they may have at the end of a measurement process.[27]

The above principles do not suffice to complete a modal interpretation. What needs to be added is a systematic treatment of the relations between values of quantities pertaining to different parts of a system. In my earliest attempt I was staying close to Everett's relative-state formalism; Nancy Cartwright (1974) showed that this failed, and that a repair I had offered failed as well. The first version that worked appeared in my "Semantic Analysis of Quantum Logic" (van Fraassen 1973) and was fully worked out in my book on quantum mechanics, eventually.[28]

26 The interpretation does still have to meet the challenge: *what is a measurement?* This has an answer within quantum theory: the physical processes that qualify for use as measurement operations can be specified as specific class of quantum-mechanical interactions. That was nothing new; a thorough quantum-mechanical theory of measurement was well developed by the time the semantic approach began, and continued to be developed throughout succeeding decades, to give precision to this subject.

27 To be even-handed, I also developed an anti-Copenhagen variant in which principle II is rejected, and all quantities can have values at all times. In this variant, a Hermitean operator will in general represent many distinct physical quantities, whose measurement outcome statistics are always the same. Michael Redhead called this, aptly, "de-occamization". Quite different versions of the modal interpretation were developed (see for example Dieks and Vermaas 1998); for a whole range of interpretations that include a variety of modal ones as well as Bohm, see Bub (1997), but that range is not exhaustive (cf. my review, van Fraassen 1998)).

28 The critique by Leeds and Healey (1996) erred by not recognizing just how values of quantities relating to different parts of a system related to each other; see my reply (van Fraassen 1997a).

3.9.4 The Quantum Mechanics Interpretation Wars: Viva Copenhagen!

So how did this Copenhagen Variant of the Modal Interpretation compare in the "interpretation wars"? While originally inspired by Everett's way of developing the theory, a true tour de force, it did not go the "multi-verse" way, in which all possibilities are realized, in alternative histories, other real worlds. Everett's way paints a picture of the universe, with its many "branching" world histories, as deterministic taken as a whole, although the appearances for us must necessarily be indeterministic. I suppose that it was a leaning to Russell's "robust sense of reality", or perhaps deeper empiricist leanings, that made it seem crucial to find an interpretation of quantum mechanics as a theory of just this actual world we live in.

Bohmian mechanics, which is another interpretation of this sort, is deterministic, and does not dabble in other world stories, but it opts for determinism without causality, so to speak: it is a non-local theory. That is not an objection, but should be kept in mind so as to deflect the impression that it restores something like the traditional causal order. It may look at first like a story about something like miniscule classical billiard balls rushing hither and thither, but really, Bohmian particles are more like moving shadows. They have position, but the story of an individual particle's position over time is its entire story. If you want to ask about mass, charge, and so on, anything that might make them look substantial, you are referred to the wave equation that choreographs the movements of those particles.

Empiricist or realist philosophical leanings cannot by any means explain advocacy of one direction or another taken in the interpretation of the physical theories that we all admire. But they do seem to play a role. As example, take the many scathing remarks that have been made about the "no-go" theorems for hidden variables, which include the much maligned proof by von Neumann that there are no dispersion-free states, and the much more important Kochen and Specker proof (Kochen and Specker 1967).

Those "no-go" arguments certainly did involve strong assumptions, and someone interested in constructing a hidden variable interpretation would have the right to reject them. Specifically, central to such no-go theorems is this assumption:

> **IV. Quantity identity:** If A and B are physical quantities, and the probability distributions for measurements of A or B in any and all states are the same, then A and B are the same quantity.

That is a strong assumption (built into the Copenhagen Variant of the Modal Interpretation). For example, it implies that a Hermitean operator can represent one, but only one, physical quantity. Both "contextuality" and hidden variables are ruled out. If there is an empiricist conviction behind this, it is this:

> **V. Norm:** If physics is to be an empirical science, then a physical theory should not draw distinctions which, according to that theory itself, make no testable difference.

To see what this rules out, imagine one kind of experiment in which a quantity *A*, say charge, is measured together with the x-coordinate of position. A second kind of experiment is just the same, except that the same apparatus is rotated, so that *A* is measured together with the y-coordinate of position.

Quantum mechanics will yield, for each preparation of state, the same probability distribution for the values of *A*, for both types of experiment. You can indeed *consistently* say, contrary to principle IV, that two different quantities were measured. That is, in any particular case, you can consistently say that, if the other kind of experiment had been done instead, a different value could or would have been found for *A*. Contextual hidden variable theories can allow for this. But you can see here how far they go, beyond what is actually delivered in a quantum-mechanical model of such an experiment, and why IV or V would be a reasonable constraint on interpretation.

4 Epistemology from within the Semantic Approach

The Received View of what theories are had had several typical consorting views concerning the epistemology of science. Confirmation theory as a refinement of the ancient idea of induction (with "instance confirmation" the new garb of numerical induction), the hypothetico-deductive method, and abduction (with its new form of inference to the best explanation, starting in the sixties with Gilbert Harman) were all presented in forms closely tied to the view of theories as sets of sentences.

How would epistemic and epistemological questions look from within a study of models and modelling?

The answers that appeared cannot be counted as part of the semantic approach, since they differed from author to author. The differences had to do to a certain extent with differences between empiricism and realism. But there was also a common theme in the rejection of traditional ideas in epistemology, and in the conviction that attention to modelling in scientific practice had to provide the key.

4.1 Elizabeth Lloyd on Darwin's Epistemology

Elizabeth Lloyd elaborated the state-space form of theory structure in her early work on population genetics, and then worked with that framework to revise received ideas in the epistemology of science, beginning with an examination of Darwin's support for his theories.

Lloyd took up this topic in her earliest articles (1983, 1984, 1986) with special attention to practices in biology.[29] Already in the first article she presented a view of Darwin's practice, in his amassing of support for the theory of evolution by mutation and natural selection, that contrasted sharply with earlier accounts. By construing Darwin's theory in the terms provided by the semantic approach, she arrived at a new way of viewing evidential support for the theory.

Lloyd's own (1984) presentation of the semantic approach emphasized those aspects that needed to be salient for this case:

> According to the semantic view of theory structure, a scientific theory specifies certain kinds of systems. The systems specified by a scientific theory are ideal; they are used in scientific explanation through claims that certain systems in the natural world are of the kind defined by the theory. The semantic view offers a formal approach to analysing these systems, which are usually understood as mathematical structures. (Lloyd 1984, 243)

She presented then, in detail, the state-space format of both deterministic and stochastic (non-relativistic) theories, illustrated throughout by applications to population genetics. What needs emphasis there is that the term "model" will typically not refer to a single model (with each parameter having a definite value) but rather a *model type*, associated with a distinctive *state space type*. Presentation of a theory will in practice specify a family of state space types, families of associated model types, and relations among them:

> In the simplest case, these models are related by having a common state space and common law of succession. Differences between the models lie in different initial conditions (i. e. initial location of the system in the state space), and hence different successions of states satisfying the same law. Generally, however, the theory allows for greater essential variety in the system it deals with (for instance, different degrees of freedom). Representation of these systems then requires models with different state spaces (for instance, of different dimensionality) and different laws. (Lloyd 1984, 262)

---

29 See further Lloyd (1988, 1994 [1988]). I regret that I must omit from this already long paper any account of the role of the semantic view in philosophy of biology, developed by Elizabeth Lloyd (1983, 1984, 1986, 1988, 1994 [1988]), Jon Beatty (1982, 1987), and Paul Thompson (1983, 1986, 1987, 1988). There is an informative, if critical, review by John D. Collier (1992).

The epistemological questions that arise now concern the way this panoply of models offers means of representation for the empirical phenomena in the theory's intended domain. However these questions are to be phrased, trying to fit them into the patterns traditionally displayed would be a truly Procrustean exercise at best.

Darwin recognized from the beginning that he could not offer direct support for his theory of the sort that was claimed, and was putatively available, for example, for Newton's theory of gravitation. But he pointed to the wave theory of light, and its mechanical ether, as not admitting such support either.

Several recent philosophy of science studies had tried to reconstruct what Darwin did offer in support, under these circumstances, in terms of the hypothetico-deductive method, of the covering law model, and of inference to the best explanation. Lloyd (1983), starting from within the semantic approach to theory structure, presented Darwin's work instead as a practice of model construction, augmented with testable empirical assumptions. This begins with an answer to a "how possible?" question:

> Particular models serve to show that the theory is compatible with observed phenomena, viz., empirical generalizations and particular facts of evolution, by providing a specific way in which the theory plus certain specific assumed conditions would produce a fit with the empirical results. (Lloyd 1983, 115)

To evaluate such an account, an empirical investigation has to target those specific assumed conditions, the "model assumptions", as well as the match of the model outcomes with data from those observed phenomena.

This evaluation will follow the representation of the theory's content in terms of models along the lines we saw above. To illustrate it, the following simple, but striking example is found in Darwin's letter about wingless beetles. There he poses a problem, offers a model that would lead to the observed phenomenon under certain testable circumstances, and then checks a variation in those circumstances:

> There is a very curious point in the astounding proportion of Coleoptera that are apterous [wingless]; and I think I have guessed the reason, viz., that powers of flight would be injurious to insects inhabiting a confined locality, and expose them to be blown to the sea: to test this, I find that the insects inhabiting the Dezerte Grande, a quite small islet, would be still more exposed to this danger, and here the proportion of apterous insects is even considerably greater than on Madeira proper. (Darwin 1919, 1: 404–405; cited after Lloyd 1983, 122)

As construed, Darwin is in effect offering two models, of Coleoptera living in two places, Madeira and Dezerte Grande, differing in the extent to which the conditions favour natural selection in one specific respect. Two assumptions need to be checked: that the two places differ in the way assumed, and that the selection mechanism is operative: the wind does blow winged beetles out to sea. Finally, the model outcome needs to match the investigated feature: there is a larger proportion of apterous Coleoptera on the small islet, where the postulated selection pressure is greater.

Witness, in this informal account, the technique for independent testing of the model assumptions. First, the crucial assumption has to be isolated from the other assumed features of the modelled situation. This is done by comparing two models belonging to a common type, which have that assumption in common, while differing in certain parameter values. In general, it is through comparison of models with other models, that a new assumption can be isolated as the one to be submitted to empirical testing. In principle this technique makes possible a rigorous, though indirect, testing of each model assumption.

It is clear that in this way of discussing a scientific inquiry, with the focus on models and modelling, the entire topic of evidential support takes on a completely different look. There is a simple empirical hypothesis involved, certainly: that the proportion of winglessness is greater on the islet. But the theory is supported by evidence confirming that empirical hypothesis only via its role in the way the theory provides its modelling technique. The data, very limited if just considered in and by themselves, are viewed "through theory-coloured glasses" so to speak.

Epistemology is no longer as neat, its problems are no longer as easily grasped, its relation to common intuitions among philosophers no longer as intimate, but it is finally in much closer touch with science in practice.

4.2 Frederick Suppe: Credentialing, and Science without Induction

The terms in this heading come directly from three of Suppe's papers (1993, 1997, 1998), that is late, really beyond the intended scope of my account. But Suppe had been talking for a long time about implications for epistemology in what we actually find in scientific papers, and there are indications to be found in his writings from the beginning. I consider these late papers, rather than his earlier work on the concept of knowledge, to be his contribution to the application of the semantic approach in epistemology.

Suppe examined a range of articles reporting experimental studies in disciplines ranging from particle physics to psychology and sociology. What he found was quite different from what someone acquainted with, for example, Carnap's or Hempel's idea of confirmation, would expect. The typical functional and argumentative structure of the papers was the following:

> They present the *reduced data* or *results of* the experiment. They make a case for the *relevance* of the experiment and its results to the concerns of the target scientific community. They provide at least some of the detail about the experimental setup or *method*—the design and circumstances of the experimental observations—needed to *replicate* or evaluate the study. They provide *an interpretation of the reduced data* (results) which yields the specific *experimental claims.* (Suppe 1997, 391–392)

What is especially noticeable is that the marshalling of evidence and the arguments offered are not so much arguments for the truth of the claim, but rather arguments that attempt to anticipate and erase specific doubts that could be raised against the proffered interpretation. While doubts are countered and competing interpretations are rebutted, such an analysis cannot be logically exhaustive; on the other hand, there is no explicit appeal to anything that a philosopher would recognize as inductive inference.

Suppe summarizes this with the slogan that the task addressed is not one of confirmation but of *credentialing:*

> This feature of science proves troublesome for most epistemologies and accounts of scientific knowledge. It is most readily accommodated by divorcing belief-formation processes from the evidential basis for knowledge. (Suppe 1997, 393)

I would put it somewhat differently: it is the distinct concepts of *confirmation* and *evidential support* that need a divorce. Confirmation consists in reasons that make something more likely to be true, while evidential support yields reasons for its acceptance, which is a pragmatic category. Empirical hypotheses are readily confirmed through statistical analysis of the data. Acceptance of a theory involves some beliefs, certainly, but also more pragmatic reasons, and evidential support through experimental studies will provide credentials for acceptance. Though differently phrased, this is in agreement with Suppe's critique of traditional epistemology.

4.3 The Problem of Empirical Adequacy

In the 1970s the question of how the relation of *empirical equivalence* is to be regarded, let alone defined, became important. If this notion could not be properly defined, or would mark a distinction without a difference, the range of possible doxastic or epistemic attitudes toward scientific theories would be limited.

To give but one influential example of the literature then: Lawrence Sklar argued in his *Space, Time, and Space-Time* (1974) that the question of relations between alternative geometries should properly be placed under the general rubric of empirical or observational equivalence for physical theories in general. Intuitively characterized, theories to be considered equivalent in this sense would have to be ones between which no observation can decide, because they have precisely the same implications for the observable phenomena.

So the intuition that there is difference between empirical adequacy and truth over all was salient in the literature. The problem seemed to be, however, that this intuition lacked any convincingly precise content.

4.3.1 No Distinction within the Received View

Does this idea, that distinct theories could be empirically equivalent, make sense at all?

Richard Boyd (1973) and Michael Gardner (1976), working entirely within the Received View of theory structure, argued that no distinction between empirical equivalence and full logical equivalence can be drawn. Both trade on the point (argued, as it happens, by Hilary Putnam at the 1968 Urbana Conference) that to identify the empirical import of a theory, we need to take into account the auxiliary hypotheses that may be apt in possible or potential experimental conditions. In this they see a problem for the very notion of empirical equivalence.

Thus Boyd asks how the thesis of under-determination by the facts or evidence about the observable

> [...] is supposed to be interpreted with respect to the employment of auxiliary hypotheses in the deduction of observational consequences from the theories in question. When [a claim] says that two theories have the same observational deductive consequences, what additional theories or laws, if any, are we to suppose are employed in making the relevant deductions? This question arises in view of the widely acknowledged fact that most of the important theories arising in the physical sciences have, unless other laws are employed with them as auxiliary hypotheses, no non-trivial observational consequences. This is, for example, true of any two consistent theories which are stated entirely in non-observational terms. (Boyd 1973, 3)

Boyd's questions are clearly posed within the assumption that the Received View is the correct view of what theories are, and that *sentences* in the language in which the theory is formulated can be identified as the "observational consequences" of that theory. The same is true of how Gardner poses the question:

> What is puzzling about this controversy is that so few of the participants have explicitly questioned the intelligibility of the terms in which it is carried on. To say that two theories θ and θ' are observationally equivalent is presumably to say that if Γ (Γ') contains just those theorems of θ (θ') which are directly testable by observation, then Γ and Γ' are identical (or at least inter-deducible with standard rules). And presumably the sentences which are directly test-

> able by observation are to be identified as those formulated in an observation language. (Gardner 1976, 106)

The problems that Gardner then sees for a distinction between empirical adequacy and truth all told become severe and numerous. There is first of all the, by that time commonplace, rejection of the observational/theoretical distinction.

But more telling is the consideration that Γ cannot contain all that we must take into account. As we saw Boyd pointing out: what is also relevant are the implications of the theory, for what is observable, conditional on statements about initial conditions and instruments, as auxiliary hypotheses about, for instance, the absence of unknown unknowns. Moreover, those additional statements from outside the theory could come from other theories, possibly at a much later time. Here Gardner cites the example made familiar by Poincaré that a hypothesis of absolute motion may have testable implications once Newton's mechanics is joined by nineteenth-century electromagnetism.

Boyd offers a destructive trilemma.

First, if any auxiliary hypotheses whatsoever are admitted then theories with the same "observational consequences" are just the same theory. Michael Gardner's argument to this effect, relying on the material conditional, shows this very well:

> One can justifiably go further, though, and assert that the claim is unintelligible. We saw above that making sense of observational equivalence requires making sense of the notion of a set Γ of observational consequences of θ. But if Γ is not the subset of theorems of θ which are formulated in the observation language (whatever that is), because additional assumptions may be required, then what is it? How is the class of permissible additional assumptions to be specified? [...] We certainly cannot allow any assumption whatever to be conjoined with θ, since then by conjoining '$\theta \supset \underline{S}$' we get any observational statement $\underline{S}$ as a consequence. Nor can we say that the assumption must be true or well confirmed, since if $\underline{S}$ has either property, so does '$\theta \supset \underline{S}$'. (Gardner 1976, 107)

So construed then, the unintelligibility of the distinction between empirical and logical equivalence is just a logical point.

What about other construals? Boyd then offers a *second* and a *third* alternative: that the set of auxiliary hypotheses be restricted to those currently accepted, and alternatively, to those that will eventually be accepted. But the theory might have relevant implications conditional on auxiliary hypotheses not currently accepted but true. As to what would eventually be accepted, if it makes sense at all, the same difficulty applies, since there could be much that is true but never accepted.

Gardner makes what seems to me the most telling point: trying to delimit the relevant auxiliary hypotheses "is about as promising an enterprise as solving the problem of initial conditions in the theory of counterfactuals" (Gardner 1976, 107).

4.3.2 Why the Distinction Should Nevertheless Be Intelligible to All Sides

The issue of empirical adequacy was clearly all enmeshed in the perceived problem about theoretical terms. However easily we can now understand that meshing in view of the Logical Positivist history, that was a mistake. The difference between what is and is not observable is a matter of fact, not one of language or description.

In a scientific description of the world, parts can be classified as observable and parts as unobservable. Relative to a theory, what the observable parts are like is intimately related to what the unobservable parts are, but from a logical point of view they are independent. What the unobservable part is like can be anything at all, as far as the observable is concerned, just as what Australia is like is logically, even if not physically, independent of the nature of Antarctica.

So there must be a way of explaining what it would mean to say that two theories agree on everything observable, but disagree overall. Grover Maxwell had placed an obstacle in the way of this point, with his arguments to the effect that there is no feasible distinction to be drawn between what is and what is not observable, by his widely accepted slippery slope argument about observability (Maxwell 1962). Slippery slope arguments are slippery at the best of times, but this one even more so (cf. van Fraassen 1980, 14–18). Not being able to rest on actual slippage, Maxwell introduced counterfactuals and science-fiction or fantasy examples about how things that are not in fact observable to us, as we are, could be observable if we, or other circumstances, were different. The inference from "could be observable" to "is observable" apparently seemed obvious to his readers, however invalid in form. The "able" in "observable" conferred a second "could" and two "*could's*" were apparently taken to collapse into one.[30]

In fact, despite the dismissal of the terms in which the problem tended to be posed, the question of *how theory relates to data and measurement outcomes* did not disappear. Could a theory satisfy all the criteria of adequacy pertinent to this relationship without being true overall? That is the same question, posed more

---

30 Also important, along these lines was a later critique by Musgrave (1985), equally focused on language: if "electrons are observable" would be an observational consequence then so is "electrons are not observable", so there are observable consequences about unobservable entities (vide my reply 1985, 256). This argument was later revived by Lipton and Dicken, with response by Muller and van Fraassen.

concretely, as the question whether empirical adequacy is distinct from truth overall. That question could not be avoided, as a topic to be broached, by philosophers of science of any particular persuasion.

The most promising beginning would be to look into the actual scientific practice for dealing with data, and most specifically measurement results. And that is indeed how the story had begun.

4.3.3 Starting with Suppes

Once again, it was Patrick Suppes who set the early standard for philosophical attention to the sciences quite unconstrained by the Logical Positivist tradition. I have a much treasured loose-leaf copy of Suppes' *Set-Theoretical Structures in Science* (Suppes 1967), though it is only the January 1970 print-out. It begins with a sharp critique of what Suppes calls the "standard formulation", that is, the standard Logical Positivist sketch of a theory as an axiomatized set of sentences in a first-order language supplemented by such putative devices as "coordinating definitions" or "empirical interpretations" to link those sentences to what they are intended to be about.

Then Suppes offers an account to replace that, beginning in a section entitled "Models versus Empirical Interpretations of Theories" (1–6). In this account, which is emphatically not in the form of a definition, theories are characterized as providing first of all a class of theoretical models, that are to be related to models constructed directly from the data.

Suppes was here introducing a new focus on the relationship between theoretical models and models of experiments and of data gathered in experiments. Here he presented, if still in an initial sketch, the importance of data models and the hierarchy of modelling activities that both separate and link the theoretical models to the practice of empirical inquiry.

Suppes took the class of theoretical models to have a set-theoretic definition that selects the mathematical structures which are to serve as models. But the word model does not mean simply mathematical structure, as if we were dealing with theories in pure mathematics. In the case of an empirical science all the weight falls on the second and third part: the relation of theoretical models to models of experiments and of data gathered in experiments, and the relation of the latter to the phenomena in the theory's intended domain of application.

His first example is still simple and stylized:

> Given an axiomatized theory of measurement of some empirical quantity such as mass, distance, or force, the mathematical task is to prove a representation theorem for models of the

> theory which establishes, roughly speaking, that any empirical model is isomorphic to some numerical model of the theory. The existence of this isomorphism between models justifies the application of numbers to things. We cannot literally take a number in our hands and apply it to a physical object. What we can do is show that the structure of a set of phenomena under certain empirical operations is the same as the structure of some set of numbers under arithmetical operations and relations. (Suppes 1967, 1–8)

Then we are introduced, if still in an initial sketch form, to the importance of data models and the hierarchy of modelling activities that both separate and link the theoretical models to the practice of empirical inquiry:

> The concrete experience that scientists label an experiment cannot itself be connected to a theory in any complete sense. That experience must be put through a conceptual grinder [...] Once the experience is passed through the grinder, what emerges are the experimental data in canonical form. These canonical data constitute a model of the experiment, and coordinating definitions are provided for this model of the experiment rather than for a model of the theory. [...] The assessment of the relation between the model of the experiment and some designated model of the theory is a characteristic fundamental problem of modern statistical methodology. (Suppes 1967, 1–12/1–13)

In an earlier paper, presented at the 1960 International Congress for Logic, Methodology and Philosophy of Science, Suppes used "realization" instead of "model", but made this same point both precise and concrete with specific illustrations. One illustration, a linear response learning theory developed by Estes and himself, proceeds as follows:

> A possible realization of the theory is an ordered triple $\boldsymbol{X} = < X, P, \theta >$ of the following sort. The set X is the set of all sequences of ordered pairs such that the first member of each pair is an element of some set A and the second member an element of some set B, where [...] the set A represents the two possible responses and the set B the two possible reinforcements. The function P is a probability measure on the smallest Borel field containing the field of cylinder sets of X; and θ, a real number in the interval $0 < \theta \leq 1$, is the learning parameter. (Suppes 1962, 253–254; fonts of mathematical symbols partly adjusted)

The only theoretical quantity here is the learning parameter, and it is easy enough to see what will be the relevant data model to be gleaned from an actual experiment. It will consist of a set of sequences of response-reinforcement pairs. The data model will present just the relative frequencies of responses conditional on the two types of reinforcements, and will be assessed in terms of their relation to the probability measure P in models of the theory with varying values of the learning parameter.

4.3.4 Generalizing on Suppes' Assessment Relation

Suppes spends a good deal of time in this article to fill in the details for my cavalier "The data model [...] will be assessed in terms of [...]". The relative frequencies are extrapolated to probabilities, and those probabilities must in effect, in order to fit perfectly, be marginals of a probability measure in one of the theoretical models. (Perfect fit not to be expected in practice, but we need to understand that case first.) Now, this is an embedding of one, small, measure algebra in another one.

Despite the greater complexity, this general point is not essentially different from what we see in the example above that I called simple and stylized. The message to take away is about how we should conceive of the appearances that are to be saved, and of the theoretical structures that are to save them.

As usual, the 25 words or less formulation ignores the devil that is in the details. When I introduced this topic, I began with a few intuitive ways of speaking:

> It requires an explication of 'empirically adequate'. For now, I shall leave that with the preliminary explication that a theory is empirically adequate exactly if what it says about the observable things and events in this world, is true—exactly if it 'saves the phenomena'. A little more precisely: such a theory has at least one model that all the actual phenomena fit inside [...]. The whole of the next chapter will be devoted to the explication of this term, which is intimately bound up with our conception of the structure of a scientific theory. (van Fraassen 1980, 12)

The "fit inside" metaphor is what specifically needed to be explained, and that required attention to how the phenomena are represented:

> To present a theory is to specify a family of structures, its *models*; and secondly, to specify certain parts of those models (the *empirical substructures*) as candidates for the direct representation of observable phenomena. The structures which can be described in experimental and measurement reports we can call *appearances:* the theory is empirically adequate if it has some model such that all appearances are isomorphic to empirical substructures of that model. (van Fraassen 1998, 64)

Note well, at this point, precisely how Suppes' notion of a data model is equated with how the system *appears* to us in the measurement outcome represented by that data model.

The measurement in question is typically a protracted affair. In learning theory, it may involve a hundred or so runs with a hundred or so subjects. In a physics or astrophysics experiment it may involve a gigabyte or so of joint measurements of several kinematic quantities. The measurement is a complex consisting of many "small" measurements, each yielding data, to form a possibly enormous data set, which goes through the conceptual grinder before the result is formulated.

A model is a mathematical structure, and if something is said to be embeddable in a mathematical structure, and this is meant literally, then that something is also a mathematical structure. So the appearances are mathematical structures too. They are the data models which form the ultimate content of the measurement outcomes. It is only because the *appearances*, though unlike phenomena they are neither concrete nor physical, are direct *representations of phenomena*, that we can say, with a quite transparent metaphor, that in such a case *the phenomena fit inside the model.*

5 What Is a Scientific Theory?

"What is ...?" is a typical philosophical question form, and usually, in the end, it doesn't have an answer. Or, equivalently, it has many answers, none of which are adequate in every way. The value of the question lies in how we come to realize that.

5.1 From Patrick Suppes to Sneed, Stegmüller, and Giere[31]

Suppes and his collaborators had sought to reformulate the foundations of Newtonian mechanics, by replacing Newton's axioms with the definition of a Newtonian mechanical system. This gives us, by example, a format for scientific theories. But a definition does not contain any information about the world, while theories do or are meant to do.

Ronald Giere's encapsulation (Giere 1979, 80–81) may be seen as starting with Suppes' format and filling that gap. In his view, a theory consists of (a) the *theoretical definition*, which defines a certain class of systems; (b) a *theoretical hypothesis*, which asserts certain relations between (sorts of) real systems and members of that class.

In the simplest case, the hypothesis could just be that certain real systems belong to (are identical with certain members of) that defined class of systems. To illustrate Giere's distinction, we can say that the hypotheses that Newton added

31 Ronald Giere attended Suppe's symposium in 1969, and was there on the night that the Received View reportedly died.

to his theoretical definitions included, for example, that our solar system is a Newtonian system, with its centre of mass at absolute rest.

There are two ways to see beyond the simplest case. One way was signalled by Joseph Sneed, on behalf of Stegmüller and himself:

> We adopted roughly the following concept of scientific theory. A scientific theory is a conceptual structure that can generate a variety of empirical claims about a loosely specified, but not completely unspecified, range of applications. (Sneed 1976, 120)

This acknowledges that some pre-theoretic identification of the target phenomena is needed, but it cannot be as precise as the retrospective characterization of those phenomena by the theory. The other way was emphasized by Giere: the relation asserted between real systems and models will be in general one of similarity rather than identity (a theme continued by Paul Teller, see e.g. Teller 2001).

Especially when "larger" philosophical issues are at stake, a general formulation may be the best vehicle for discourse or debate. Generalization can illuminate, but it can also obscure. Thus, a neat general formulation such as Giere's, which was very useful, could also lead to misunderstanding.

Thus Stathis Psillos dismisses the semantic view in Giere's formulation with:

> Be that as it may, what is noteworthy is that the theoretical hypotheses do all the representational work that has to be done if theories are to acquire empirical content. In fact, theoretical hypotheses are linguistic entities which have truth-values. (Psillos 1999, 274)

Psillos adds a bit later: "This being true, the difference is not really important" (Psillos 1999, 274), the difference in question being that between the Received View and the semantic view.

As Psillos reads Giere, therefore, theories are really sets of theoretical hypotheses, though formulated as describing a relation to models. That may seem plausible at the very abstract, idealized level where Giere sums up that relation as a similarity.

However, when we look at specific scientific specifications of such relations, we do not find anything like the axiomatic systems envisaged as rational reconstructions of scientific theories envisaged in the Received View. First of all, we find ways in which physical quantities are represented within the theory. Thus at the outset it is specified in classical mechanics that physical quantities are represented by infinitely differentiable vector-valued functions, in general relativity by tensors, in quantum mechanics by Hermitean operators. Then the empirical content appears when it is asserted that the measurement results are accommodated when the state assigned to an object is, for example, a quantum state rather than a classical mechanical state.

Could all this in principle be formulated as a description of the object in a language with terms for the physical magnitudes in question, in which the basic principles appear as axioms describing that sort of object? We could look carefully at some limited attempts to do so that have appeared in the literature, but the scepticism with respect to any such enterprise on a realistic level, that had increased rapidly starting in the sixties and that motivate the semantic approach, had considerable warrant.[32] But it would be more instructive to look to the many concrete examples of scientific theorizing that Giere offers in his books (Giere 1979, 1988).

5.2 Taking a Wider View

A few times in the literature there appeared a caricature of the semantic view of theories: "a theory is a set of models". That this was the view of theories within the semantic approach could be backed up by convenient quotation mentioned in a footnote above:

> If the theory as such, is to be identified with anything at all—if theories are to be reified—then a theory should be identified with its class of models. (van Fraassen 1989, 222)

But that sentence was part of a brief exposition of Patrick Suppes' approach in the fifties, a precursor to the semantic view. Here "models" has the sense it has in Suppes' presentation: "When a theory is axiomatized by defining a set-theoretical predicate, by a *model* for the theory we mean simply an entity which satisfies the predicate" (Suppes 1957, 253). That a theory is to be identified with its class of models was not part of the later semantic view. Though that class plays a prominent role, it is accompanied by what makes the theory an empirical theory rather than a bit of mathematics:

> From this point of view, the essential job of a scientific theory is to provide us with a family of models, to be used for the representation of empirical phenomena. On the one hand, the theory defines its own subject matter—the kind of systems that realize the theory; on the other hand, empirical assertions made by holders of the theory have a single form: the phenomena can be represented by the models provided. (van Fraassen 1972, 310)

---

32 In addition to what I already described above, attention would have to be paid to serious questions about the limits to axiomatizability and definability. The best examples to be found are axiomatizations of classical mechanics, but these rely on reference to parts of classical mathematics that are not finitely or recursively axiomatizable.

When it became imperative to formulate the semantic view in ways not explicitly wedded to the format of theoretical physics, I accepted Giere's view in the main. Because of my concern to develop a clearly empiricist view, however, I offered a somewhat more specific form for both the theoretical definition and hypothesis. Models have specified substructures, their *empirical substructures*, and the hypotheses are claims that all the relevant phenomena were correctly represented by the empirical substructures in one of those models.

5.3 Theories and the Phenomena: Apropos Halvorson's Critique

In the passage just quoted, the word "representation" is salient, in what it asserts about the models. Models are mathematical structures, but they are structures used for representation, and it is this function, of representing the empirical phenomena, that makes them scientific models.

To bring out the diverse aspects of this intricate relation between theories, models, and phenomena, we can take a look at the recent critique of the semantic approach by Hans Halvorson (2012). This was also the occasion of Halvorson's announcement of an exciting new program in formal philosophy of science, which continues some aspects of the semantic view. But here I will limit discussion to the critique, in order to shed light on what a scientific theory is and how theories are related to classes of models (see further van Fraassen 2014).

We can begin at once with what Halvorson announces as the starting point of his critique:

> For my argument to go through, I only need the semanticist to grant a weak sufficient condition on theory-hood: the class Mod(T) of models of a first-order theory T is (the mathematical part of a) theory in their sense. (Halvorson 2012, 189)

That passage is quite striking: *that*, I would say, is how the semantic approach must have looked from within the Received View. Imagine a philosopher, one for whom a first-order theory is the paradigm notion of theory, grappling with the idea that attention should be shifted from sentences to models. What could she imagine but that the idea is to think solely about Mod(T) without its relation to T? And how exceedingly puzzling, even incoherent, that must have seemed!

A look at the notion of equivalence of theories, as addressed by Halvorson and as it appears in the history of science, will bring out both the place of the way Halvorson treats such a subject and what needs to be added to illuminate actual scientific practice

5.4 Equivalence and Non-equivalence of Scientific Theories

When it comes to a notion of equivalence of theories, Halvorson offers three criteria, formulated so as to refer only to models, in terms of mathematical structure without reference to any language of which they would be models. For that is what he takes to be required given the semantic approach's rejection of the Received (syntactic) View of theories (as witnessed in the above quotation). Those three criteria are naïve even at first blush, and it is no surprise that they do not in any way correspond to definitional equivalence of first-order theories. Thus, if the passage just quoted is a good basis for a critique of the semantic view then this shows pretty well that on the semantic view there is no good criterion of equivalence for theories.

Halvorson is right to this extent: criteria of equivalence which refer to the members of such a class of models, with no reference to their connection with language, do not in any way correspond to any significant equivalence relation between first-order theories. But note well: such criteria also do not refer to the representational character of scientific models, nor to the theoretical hypotheses concerning their relation to the phenomena.[33] Therefore these arguments do not touch on much of the semantic approach at all, they isolate something which leaves us separated from the crucial difference between scientific theories and pure mathematics.

However, Halvorson's conclusions will help us to highlight some interesting features of theories, as understood within the semantic approach, though *malgré lui*, for those are features he understands quite differently. I mean his conclusions that theories may be the same although the corresponding sets of models are different, and conversely that theories may be different although the corresponding sets of models are the same. If we take the corresponding sets of models to be what a scientist would present as such (instead of an elementary class of models of the sort Halvorson considers), these conclusions are certainly correct.

First, the set of mathematical structures that satisfy an equation may be the same when that equation is offered to describe one sort of phenomenon as when it is offered to describe another sort. As an example, take the simple diffusion equation:

$$\frac{\partial \varphi(r,t)}{\partial t} = D\nabla^2\varphi(r,t)$$

33 In his second footnote Halvorson (2012) points out that he is ignoring that component, but does not see that as affecting his critique.

If one scientist presents it to describe gas diffusion and another scientist presents it to describe temperature variation over time, they are presenting us with the same set of mathematical structures, that is, the set of solutions of this equation. But are they presenting the same theory? Clearly not, since one offers these models as representations of gas diffusion, and the other offers them as representations of thermal diffusion. Being theories *of* different phenomena, though representing both *as* that sort of process, they are clearly different theories, *though the sets of models are the same.*

What about the contrary case, of two quite different sets of models, though the theory is the same? An example comes from twentieth-century efforts to present classical mechanics in a more precise and rigorous form. In most systems of moving bodies, if the system is assumed to be a Newtonian system, the kinematic data are sufficient to determine the masses. But there are exceptions. The extreme case is a model in which there is only one body, subject to zero total force, and never accelerated.

Here there were several alternatives, though they would probably not have occurred to Newton as being of any consequence. One obvious suggestion would be to add just the principle that everything always has a determinate mass, and this mass does not change if the impressed force is zero. An unaccelerated body would then be assigned some mass in each model, with no limits on how large a mass, beyond what the kinematic data could rule out. That would introduce many distinct models that would be the same in all kinematic respects, and thus candidates for representation of all the same phenomena. Another suggestion, implemented actually by Herbert Simon (1954), is to leave the mass undefined in such cases: mass is then a partial function.

We can see the difference, and if any difference makes a difference then these are different theories, and different from the original mechanics. But if we have the details of any model of the one sort, we can determine exactly which model or models of the other sort correspond to it exactly, and save the same phenomena. That is a simple equivalence relation, and so natural that it makes more sense to say that, although the classes of models are not the same, we have here two versions of the same theory.

Note, however, in both examples, the important role of the reference to the theory's relation to the empirical phenomena. That this relation is actually paramount in scientific discussion of theory equivalence appears clearly, for example, in the history of the question of whether the Schrödinger and Heisenberg formulations of quantum mechanics were, or are, equivalent.

Today the right way to approach this question is to begin with von Neumann's Uniqueness Theorem (von Neumann 1931), and to present both formulations within the Hilbert space formalism fully developed in von Neumann's book (1955

[1932]). The salient difference between the two formulations is that for Schrödinger, the quantum states but not the observables are functions of time, whereas for Heisenberg it is just the other way around. Within von Neumann's formalism, the observables in the two formulations are then related as follows:

$$A_{HEISENBERG}(t) = U^{\dagger}(t,0)A_{SCHRÖDINGER}U(t,0)$$

The result follows very quickly then that the different calculations for measurement outcomes are matched and yield the same expectation values for the corresponding operators.

But this is a mathematical retrospective. It informs any contemporary presentation, but was presented by von Neumann after the scientific community had already accepted that the two formulations were in any important sense "the same" theory. With typical physicists' nonchalance concerning mathematical niceties, they had accepted this on the basis of a limited result, not of mathematical equivalence as was officially claimed, but of empirical equivalence.

Briefly, it went as follows (for the full story see Muller 1997).[34] In response to the successes of Heisenberg's matrix mechanics Schrödinger published a result (Schrödinger 1926), by which he explained the apparent empirical equivalence between the two formulations, namely that from a "formal mathematical standpoint", the two theories were identical. In fact, however, his proof did not establish that at all, nor could it have, for as they stood at that time, the theories were not mathematically equivalent. Yet the physics community quickly adapted to the way in which Schrödinger had related the two theories to each other, and did not treat them as distinct alternatives.

What was actually established by Schrödinger? That his wave mechanics and Heisenberg's matrix mechanics were empirically equivalent in a limited domain, namely for the empirical phenomena (intensity and frequency of emitted or absorbed radiation) for which at that time they could claim success. The result that actually mattered for the scientific community was the relation these two apparently very different quantum theories bore to the known empirical phenomena.

34 Muller was working explicitly within Suppes' approach to theory structure, which he developed further in Muller (1998); see further Muller (2011). As he presents Schrödinger's results, they concerned the embeddability of empirical substructures in theoretical models.

5.5 Reconsidering the Question "What Is a Scientific Theory?"

Quite possibly Halvorson would be less than satisfied with these sorts of reflections. For he is concerned to have a straightforward answer to the question "what is a scientific theory?" The informality of the semantic approach leaves, from that point of view, much to be desired. In response I would caution that a question determines its own set of possible direct answers. That is a good reason why we should be a bit wary of answering a question directly.

A scientific theory is something that can be believed or disbelieved, doubted, entertained, accepted or rejected. None of that is possible unless a theory gives some contentful information about what things are like. But should we single out a particular thing that the theory itself is? It may be better to focus on function than on form (van Fraassen 1972, 310; 1980, 64).

My two examples above, of how theories may be the same or different while the sets of models are different or the same, crucially involved the notion of representation. To present a theory is to present us with a set of models as candidates for the representation of a certain kind of physical system, or of a certain range of phenomena. We can understand that, I think, without reifying a theory as a sort of thing. But if one were to insist on the need to reify, then, yes, I think we'd have to say that a theory is a set of models. But think of parallel examples. What is a teacup? If we have to identify the teacups in this world, then the identification will be with a set of physical objects of a certain shape etc.; each teacup is one such object. But it is a teacup *because of how it is used.* A scientific model is a mathematical structure, but it is a model because it is a structure that is used (or offered for use, or contemplated as usable) in a certain way.

The better way to reflect on this is to ask how exactly theories appear on the scientific stage. A theory is presented in the first place as

> something that can be believed, doubted, accepted, disbelieved, ...
> *hence,*
> something that gives information about what things are like
> *hence,*
>> if it consists in offering models, they must be being offered as representations of how things are.
>
> *So it would seem best to conclude that*
> a theory is to be identified by "of" and "as": representation *of* ... *as* —.

For example,
- the wave theory of light represents light as a wave motion in a material medium,
- the kinetic theory of heat represents a gas as an ensemble of molecules,

and if this does not engender a definition to the rigorous mind's liking, at least it stays close to what we see in practice.

6 The End of the Affair (?)

In the 1990s the philosophy of science literature and conference scene was rapidly becoming awash in studies of models and modelling. This was due not only to the increasingly prevalent semantic approach but also to two other developments. One was structuralism as a view of theory structure using set theory along the lines of Patrick Suppes, developed by Joseph Sneed, Wolfgang Stegmüller, C. Ulises Moulines, and their colleagues (see e.g. Sneed 1979; Moulines 1975). This approach was marked by its high degree of technical development. The other, angling sharply away from formal or logic-oriented approaches to science, was initiated by Nancy Cartwright, who moved the discussion away from high theory, closer to experimentation and modelling in practice.[35] The decade ended in a cornucopia of studies of models as mediators, models as instruments, simulations as models, models as tools, and modelling as paradigm of scientific representation.

So what was arriving in this decade: a continuation, a sequel, or a replacement for the semantic approach? To end the account, I will briefly describe one new strand in the story, limiting myself mainly to the nineties.

6.1 R. I. G. Hughes View of the Situation (1996)

One definite view about the development of the semantic view was presented by R. I. G. Hughes, who initiated the debates about scientific representation that were shortly to follow. Hughes had placed his book, *The Structure and Interpretation of Quantum Mechanics*, squarely in the semantic approach (cf. Hughes 1989, 257), and in 1996 he published an article in the *Encyclopedia of Applied Physics* to describe that approach.

In this Encyclopedia article the development of the semantic approach is presented as having had three main stages. Although Hughes begins by crediting the origin of the approach to Evert Beth, he takes as Stage One the work by Suppes and his collaborators in the fifties and sixties, as continued by Sneed and Stegmüller.

35 For relations between Nancy Cartwright's approach and the semantic approach see Peschard and van Fraassen (2018, 28–29) and van Fraassen (2008, 310–311).

As Stage Two he lists the state-space approach described above, which he sees as implicitly illustrated previously in the work of von Neumann and Mackey, and later in his own, among others. His exposition here is to my mind the best short account of the state-space version to be found. He comments that the state-space representation of quantum theory leads naturally to the "quantum logical" approach to its interpretation (Hughes 1996, 178).

So far the story is essentially similar to how I told it. But then Hughes introduces what he sees as Stage Three in the development of the semantic view, under the name of the Representational Approach, and adds "of the three versions of the semantic view examined, it is the one that most nearly conforms to scientific practice" (Hughes 1996, 178).

Representation is in itself a topic of its own, with much debate about its role and aim in art, literature, and philosophy of mind. For representation in the sciences, the paradigm case is the model, and we can preview Hughes account by noting a crucial point about directionality:

> Models *look in two directions at once:*
> a model is a model *of* a phenomenon (it *represents* the phenomenon)
> *and*
> a model is a model *of* a theory (it *satisfies* the equations the theory gives us).

Both aspects of modelling need to be studied to arrive at a balanced view. If we neglect the first, we will be restricted to pure meta-mathematics, and lose touch with the specific character of the empirical sciences. If we neglect the second, we'll land in puzzlement with Margaret Morrison's apt question: "Where have all the theories gone?" (Morrison 2007).

Hughes introduced representation by highlighting a problem that he regarded as not addressed in the foregoing:

> The set-theoretic and state-space versions of the semantic account emphasize foundational theories of physics, like classical mechanics and quantum mechanics. They offer alternative ways of representing the internal structure of these theories. They do not examine the question of how a theory is applied. In contrast, the representational approach accommodates both foundational theories and their applications. (Hughes 1996, 178)

There are two salient characteristic differences between what he sees as this new stage and the preceding: (1) a new focus on "local" models of specific phenomena or sorts of phenomena, as opposed to the whole class of models of a theory, and (2) emphasis on *representation* as the concept or rubric under which this modelling activity is to be understood.

6.2 Representation as Focal Concept

Hughes sees this new stage as beginning in the 1980s, citing examples from Ronald Giere (1984, 1985, 1988), Frederick Suppe (1989), and especially Nancy Cartwright (1983). And indeed, the radical turn to local modelling had come with Nancy Cartwright's *How the Laws of Physics Lie*, who placed specific models of particular parts of the world, especially in applications of physics and in experimental work, on central stage.

As to representation, we can quite clearly see its relation to the earlier stages, since scientific models were from the outset conceived of as candidate-representations. For example:

> A physical theory then typically uses a mathematical model to represent the behavior of a certain kind of physical system. (van Fraassen 1970, 328)

What was not evident in those earlier stages was any concern to explicate the pertinent concept of representation itself, despite the frequent occurrence of that term and its cognates.

Both the terms "representation" and "interpretation" had been adapted in mathematical and meta-mathematical terminology, in ways that left them shorn of much of their natural content. This, we can see in retrospect, hampered the philosophical discussion; too often there was little or no attempt to distinguish either of them from such simpler notions as likeness, resemblance, or isomorphism.

It was mainly in connection with the philosophy of art and of literature that their meaning had received much attention. Nevertheless, through the work of Nelson Goodman and Catherine Elgin, the insights gained there seeped into general epistemology and philosophy of science. One thing to be noticed was that representation and interpretation are, as it were, two sides of one coin. Both representation and interpretation come with an "of" and an "as", and are as it were each other's dual.[36]

The first publication properly rectifying philosophy of science's lack of attention to these concepts is the article "Models and Representation" that Hughes presented that same year at the Philosophy of Science Association (Hughes 1997). This is now recognized as a seminal paper in the area, inspirational to both lovers and detractors of the semantic approach. And the literature on representation, within

36 As in, for example, "the Mona Lisa is a representation of the Gioconda as inscrutable", which is one interpretation that can be offered of that painting's content. Van Fraassen and Sigman (1993) explores some of these connections between interpretation and representation in science and in the arts.

philosophy of science, blossomed in the decade following, with notable contributions especially by Ronald Giere (1999, 2008, 2010), Paul Teller (2008), van Fraassen (2008), and Mauricio Suárez (1999, 2003, 2010).

The concept of representation is as slippery, Hughes writes, as the concept of model itself, and scientific representation takes many forms.

> What, we may ask, does the representation of a crystal as an arrangement of rods and spheres have in common with the representation of the motion of a falling body by the equation $s = gt^2/2$? Apart, of course, from being the sort of representation that a model provides. (Hughes 1997, 325–326)

But then Hughes uses the second example, as first studied by Galileo, as guide to arrive at his account. In the Third Day of *Discourses Concerning Two New Sciences* Galileo draws several diagrams to represent bodies, one in accelerated motion starting from rest, and the other in uniform motion, for the same duration. He demonstrates by their means that the two distances travelled will be equal exactly if the final speed of the accelerating body is twice the uniform speed of the other.

The word "demonstrate" is crucial here: something was used to represent those motions, and the representation was used to demonstrate something about how they compare. We inherit the word from the Latin "monstrare", *to show.* The demonstration succeeds because the representing diagrams show the relation between those two motions. But of course they only do so because the diagrams were, first of all, presented to serve as representations *of* the motions, to represent them *as* having a certain character or form, so that relationships within the diagrams could be taken (read "interpreted") to mirror certain relationships among those motions.

Hughes calls his generalization of this example the DDI account: "DDI" for "Denotation-Demonstration-Interpretation". The word "of" is glossed with denotation; the crucial word "as" has "I" as its counterpart in the acronym. The second "D" plays a central role in Hughes account: a model is something that has a use, or rather, is something that is used in several ways. It is used to "stand for" the phenomenon under study, it is used to depict it as being of a certain sort—e.g. to depict light as a wave motion, to depict an atom as like a solar system, to depict a gas as a swarm of billiard balls. When so used, it can be used as basis for a demonstration about that phenomenon.

Notice that the entire focus in this account is on the relationship-in-use between the model and its target, as opposed to a relation between the model and the theory that governed its construction. Attention has simply shifted, but the context is by no means removed. Background theory is just not evident to us at the stage where we see Galileo at work, unfamiliar as we are with the scientific background to Galileo's thinking. That may be why Hughes's example works so well for us, for he can present it, so to speak, a-historically. In the more generally familiar cases, that is not so, but on the other hand, the distance between "foundational" theories and local application, emphasized by Cartwright, quickly springs to light:

> In order to apply a foundational theory to a particular physical system we need to model the system in terms of the configurations amenable to theoretical treatment. This procedure may itself require considerable ingenuity, even if the physical system is comparatively simple. Notoriously, if we model a 3-body system within Newtonian celestial mechanics, analytic solutions of the relevant equations are unobtainable. Newton's strategy in treating the solar system is, first, to represent it as an abstract system of non-interacting bodies in a centripetal field, and subsequently to adjust this model to allow for perturbations due to interactions between selected pairs of bodies. (Hughes 1997, 333–334)

Thus we must indeed think of models as, so to speak, looking in two directions, in the way I noted at the outset.

6.3 Appreciation, and a Look Forward

As I write in 2018, the semantic approach is about 50 years old, just as the Received View was when it (putatively) died. Undoubtedly the semantic approach too is already supplanted or in the process of being supplanted. Soon now someone may cite a conference whose opening night was the night it died!

Then of course some more history will repeat itself. In recent years such writers as Michael Friedman and Alan Richardson have shown us how much valuable and truly revolutionary insight into scientific theory and practice the Logical Positivists brought us, and how shallow are the caricatures that we criticized in classes and lectures. So, some years from now there will be, I hope, some new writers who will not rue the passing of the semantic approach, but show similarly how valuable it was after all, and how shallow the caricatures that people criticized.

So, is a new dominant view of scientific theory structure already in the offing? Will it be one of the new directions I sketched here:

- Philosophy of science in practice?
- Science as representation and perspective, more broadly conceived?
- A new formal philosophy of science, drawing on contemporary logic and model theory?

Or was my sketch myopic, is a quite different, entirely new philosophy of science already standing in the coulisses, about to go on stage?

References

Beatty, J. 1982. What's Wrong with the Received View of Evolutionary Biology? In: *PSA: Proceedings of the Biennial Meeting of the Philosophy of Science Association* 1980, Vol. 2: 397–426.

Beatty, J. 1987. On Behalf of the Semantic View. *Biology and Philosophy* 2: 17–23.

Becker, A. 2018. *What Is Real? The Unfinished Quest for the Meaning of Quantum Physics.* New York: Basic Books.

Beth, E. W. 1948. *Natuurphilosophie.* Gorinchem: Noorduyn.

Beth, E. W. 1948/1949. Analyse Sémantique des Théories Physiques. *Synthese* 7: 206–207.

Beth, E. W. 1949. Towards an Up-to-Date Philosophy of the Natural Sciences. *Methodos* 1: 178–185.

Beth, E. W. 1960. Semantics of Physical Theories. *Synthese* 12: 172–175.

Beth, E. W. 1963. Carnap's Views on the Advantages of Constructed Systems over Natural Languages in the Philosophy of Science. In: *The Philosophy of Rudolf Carnap*, ed. P. A. Schilpp, 469–502. Peru, IL: Open Court.

Birkhoff, G. and von Neumann, J. 1936. The Logic of Quantum Mechanics. *Annals of Mathematics*, Second Series, 37: 823–843.

Boyd, R. N. 1973. Realism, Underdetermination and a Causal Theory of Evidence. *Nous* 7: 1–12.

Bub, J. 1997. *Interpreting the Quantum World.* Cambridge: Cambridge University Press.

Cartwright, N. D. 1971. A Dilemma for the Traditional Interpretation of Quantum Mixtures. In: *Proceedings of the Biennial Meeting of the Philosophy of Science Association*, eds. K. Schaffner and R. S. Cohen, 251–258. San Diego, CA: Philosophy of Science Association.

Cartwright, N. D. 1974. Van Fraassen's Modal Model of Quantum Mechanics. *Philosophy of Science* 41: 199–202.

Cartwright, N. D. 1983. *How the Laws of Physics Lie.* Oxford: Clarendon Press.

Church, A. 1937. Review of Garrett Birkhoff and John von Neumann, The Logic of Quantum Mechanics. *Journal of Symbolic Logic* 2(1): 44–45.

Churchland, P. M. and Hooker, C. A. (eds.). 1985. *Images of Science. Essays on Realism and Empiricism, with a Reply from B. C. van Fraassen.* Chicago: University of Chicago Press.

Clark, J. T. S. J. 1983 [1966]. Science and Some Other Components of Intellectual Culture. In: *Mind and Cosmos*, ed. R. G. Colodny, 292–307. Lanham, MD: University Press of America.

Collier, J. D. 1992. Critical Notice: Paul Thompson. *The Structure of Biological Theories. Canadian Journal of Philosophy* 22: 287–298.

Darwin, C. 1919. *Life and Letters of Charles Darwin*, ed. Frances Darwin. New York: D. Appleton.

Demopoulos, W. 2003. On the Rational Reconstruction of Our Theoretical Knowledge. *British Journal for the Philosophy of Science* 54: 371–403.

Dieks, D. and Vermaas P. E. (eds.). 1998. *The Modal Interpretation of Quantum Mechanics.* Dordrecht: Springer.

Feyerabend, P. K. 1981 [1957]. An Attempt at a Realistic Interpretation of Experience. In: *Realism, Rationalism and Scientific Method: Philosophical Papers Vol. 1*, 17–36, Cambridge: Cambridge University Press. [originally in *Proceedings of the Aristotelian Society* 58: 143–157].

Frigg, R. 2006. Scientific Representation and the Semantic View of Theories. *Theoria* 55: 37–53.

Gardner, M. 1976. The Unintelligibility of "Observational Equivalence". In: *PSA: Proceedings of the Biennial Meeting of the Philosophy of Science Association* 1976, Vol. 1: 104–116.

Giere, R. N. 1979. *Understanding Scientific Reasoning.* New York: Holt, Rinehart, and Winston. Second edition 1984.

Giere, R. N. 1984. Toward a Unified Theory of Science. In: *Science and Reality*, eds. J. T. Cushing, C. F. Delaney and G. Gutting, 5–31. Notre Dame: University of Notre Dame Press.

Giere, R. N. 1985. Constructive Realism. In: Churchland and Hooker (1985), 75–98.

Giere, R. N. 1988. *Explaining Science: A Cognitive Approach.* Chicago: University of Chicago Press.

Giere, R. N. 1999. Using Models to Represent Reality. In: *Model-Based Reasoning in Scientific Discovery*, eds. L. Magnani, N. J. Nersessian and P. Thagard, 41–57. New York: Kluwer/Plenum.

Giere, R. N. 2008. Models, Metaphysics and Methodology. In: *Nancy Cartwright's Philosophy of Science*, eds. S. Hartmann and L. Bovens, 123–133. New York: Routledge.

Giere, R. N. 2010. An Agent-Based Conception of Models and Scientific Representation. *Synthese* 172: 269–281.

Halvorson, H. 2012. What Scientific Theories Could Not Be. *Philosophy of Science* 79: 183–206.

Hanson, N. R. 1958. *Patterns of Discovery: An Inquiry into the Conceptual Foundations of Science.* Cambridge: Cambridge University Press.

Hughes, R. I. G. 1989. *The Structure and Interpretation of Quantum Mechanics* Cambridge: Harvard University Press.

Hughes, R. I. G. 1996. Semantic View of Theories. In: *Encyclopedia of Applied Physics. Vol. 17*, eds. G. L. Trigg and E. H. Immergut, 175–180. Weinheim: VCH.

Hughes, R. I. G. 1997. Models and Representation. *Philosophy of Science* 64: 325–336. Reprinted as Ch. 5 of Hughes (2010).

Hughes, R. I. G. 2010. *The Theoretical Practices of Physics.* Oxford: Oxford University Press.

Kochen, S. and Sprecker, E. P. 1967. The Problem of Hidden Variables in Quantum Mechanics. *Journal of Mathematics and Mechanics* 17: 59–87.

Kuhn, T. S. 1962. *The Structure of Scientific Revolutions.* International Encyclopedia of Unified Science, Vol. 1 and 2, Foundations of the Unity of Science, Vol. 2, No. 2. Chicago: University of Chicago.

Kuhn, T. S. 1993. Introduction. *PSA: Proceedings of the Biennial Meeting of the Philosophy of Science Association* 1992, Vol. 2: 3–5.

Leeds, S. and Healy, R. 1996. A Note on van Fraassen's Modal Interpretation of Quantum Mechanics. *Philosophy of Science* 63: 91–104.

Lewis, D. K. 1970. How to Define Theoretical Terms. *Journal of Philosophy* 67: 427–446.

Lewis, D. K. 1983. New Work for a Theory of Universals. *Australasian Journal of Philosophy* 61: 343–377.

Lloyd, E. A. 1983. The Nature of Darwin's Support for the Theory of Natural Selection. *Philosophy of Science* 50: 112–129.

Lloyd, E. A. 1984. A Semantic Approach to the Structure of Population Genetics. *Philosophy of Science* 51: 242–264.

Lloyd, E. A. 1986. Thinking about Models in Evolutionary Theory. *Philosophica* 37: 87–100.

Lloyd, E. A. 1988. The Semantic Approach and Its Application to Evolutionary Theory. *PSA: Proceedings of the Biennial Meeting of the Philosophy of Science Association* 1988, Vol. 2: 278–285.

Lloyd, E. A. 1994 [1988]. *The Structure and Confirmation of Evolutionary Theory.* 2nd ed. Princeton: Princeton University Press.

Maxwell, G. 1962. The Ontological Status of Theoretical Entities. In: *Scientific Explanation, Space, and Time: Minnesota Studies in the Philosophy of Science*, Vol. 3, eds. Herbert Feigl and Grover Maxwell, 181–192. Minneapolis: University of Minnesota Press.

Montague, R. 1962. Deterministic Theories. In: *Decisions, Values, and Groups*, ed. E. Willner, 325–370. New York: Pergamon Press. Reprinted as Ch. 11 of Montague (1974), 303–359.

Montague, R. 1974. *Formal Philosophy: Selected Papers of Richard Montague*, ed. R. H. Thomason. New Haven: Yale University Press.

Morrison, M. 2007. Where Have All the Theories Gone? *Philosophy of Science* 74: 195–228.

Moulines, C. U. 1975. Review of Sneed 1971. *Erkenntnis* 9: 423–436.

Muller, F. A. 1997. The Equivalence Myth of Quantum Mechanics. Parts I and II. *Studies in the History and Philosophy of Modern Physics* 28(1–2): 35–61 and 219–247.

Muller, F. A. 1998. *Structures for Everyone: Proofs and Contemplations in the Foundations and Philosophy of Physics and Mathematics.* Amsterdam: Gerits & Son.

Muller, F. A. 2011. Reflections on the Revolution at Stanford. *Synthese* 183: 87–114.

Musgrave, A. 1985. Constructive Empiricism and Realism. In: Churchland and Hooker (1985), 196–208.

Peebles, P. J. E. 1992. *Quantum Mechanics.* Princeton: Princeton University Press.

Peschard, I. F. and van Fraassen, B. C. (eds.). 2018. *The Experimental Side of Modeling.* Minneapolis: University of Minnesota Press.

Psillos, S. 2010. *Scientific Realism: How Science Tracks Truth.* London: Routledge.

Putnam, H. 1962. What Theories Are Not. In: *Logic, Methodology and Philosophy of Science*, eds. E. Nagel, P. Suppes and A. Tarski, 240–251. Stanford: Stanford University Press.

Putnam, H. 1974. Scientific Explanation. An Editorial Summary-Abstract. In: Suppe (1974), 424–433.

Putnam, H. 1977. Realism and Reason. Presidential Address to the Eastern Division of the American Philosophical Association, December 1976. *Proceedings and Addresses of the American Philosophical Association* 50: 483–498.

Putnam, H. 1980. Models and Reality. Presidential Address to the Association for Symbolic Logic, 29 December 1977. *The Journal of Symbolic Logic* 45: 464–482.

Rovelli, C. 1996. Relational Quantum Mechanics. *International Journal of Theoretical Physics* 35, 1637–1678.

Rovelli, C. 2017. Space Is Blue and Birds Fly Through It. *Philosophical Transactions of the Royal Society A* 376: 2017.0312.

Schilpp, P. A. 1963. *The Philosophy of Rudolf Carnap.* La Salle, IL: Open Court.

Schrödinger, E. 1926. Über das Verhältnis der Heisenberg-Born-Jordanschen Quantenmechanik zu der meinem. *Annalen der Physik* 384: 734–756. https://doi.org/10.1002/andp.19263840804 (last accessed: 7. 9. 2022).

Simon, H. 1954. The Axiomatizations of Classical Mechanics. *Philosophy of Science* 21, 340–343.

Sklar, L. 1974. *Space, Time, and Spacetime.* Berkeley: University of California Press.

Sneed, J. D. 1976. Philosophical Problems in the Empirical Science of Science: A Formal Approach. *Erkenntnis* 10: 115–146.

Sneed, J. D. 1979. *The Logical Structure of Mathematical Physics*, 2nd ed. Dordrecht: Reidel.

Suárez, M. 1999. Theories, Models and Representations. In: *Model-Based Reasoning in Scientific Discovery*, eds. L. Magnani, N. Nersessian and P. Thagard, 75–83. New York: Plenum.

Suárez, M. 2003. Scientific Representation: Against Similarity and Isomorphism. *International Studies in the Philosophy of Science* 17: 225–244.

Suárez, M. 2010. Scientific Representation. *Philosophy Compass* 5: 91–101.

Suppe, F. R. 1967. *Meaning and Use of Models in Mathematics and the Exact Sciences.* Dissertation, University of Michigan.

Suppe, F. R. 1972. What's Wrong with the Received View on the Structure of Scientific Theories? *Philosophy of Science* 39: 1–19.

Suppe, F. R. 1974. *The Structure of Scientific Theories.* Urbana, IL: University of Illinois Press.

Suppe, F. R. 1989. *The Semantic Conception of Theories and Scientific Realism.* Urbana, IL: University of Illinois Press.

Suppe, F. R. 1993. Credentialing Scientific Claims. *Perspectives on Science* 1: 153–203.

Suppe, F. R. 1997. Science without Induction. In: *The Cosmos of Science: Philosophical Problems of the Internal and External Worlds*, eds. J. Earman and J. D. Norton, 386–429. Pittsburgh: University of Pittsburgh Press.

Suppe, F. R. 1998. The Structure of a Scientific Paper. *Philosophy of Science* 65: 381–405.

Suppe, F. R. 2000. Understanding Scientific Theories: An Assessment of Developments, 1969–1998. *Philosophy of Science* 67: 102–115.

Suppes, P. 1957. *Introduction to Logic.* New York: Van Nostrand.

Suppes, P. 1960. A Comparison of the Meaning and Uses of Models in Mathematics and the Empirical Sciences. *Synthese* 12: 287–301.

Suppes, P. 1962. Models of Data. In: *Logic, Methodology, and Philosophy of Science: Proceedings of the 1960 International Congress*, eds. E. Nagel, P. Suppes and A. Tarski, 252–261. Stanford: Stanford University Press.

Suppes, P. 1967. *Set-Theoretical Structures in Science.* Stanford: Institute for Mathematical Studies in the Social Sciences, Stanford University.

Teller, P. 2001. Twilight of the Perfect Model Model. *Erkenntnis* 55: 393–415.

Teller, P. 2008. Representation in Science. In: *The Routledge Companion to the Philosophy of Science*, eds. Stathis Psillos and Martin Curd, 435–441. New York: Routledge.

Thompson, P. 1983. The Structure of Evolutionary Theory: A Semantic Approach. *Studies in History and Philosophy of Science* 14: 215–229.

Thompson, P. 1986. The Interaction of Theories and the Semantic Conception of Evolutionary Theory. *Philosophica* 37: 28–37.

Thompson, P. 1987. A Defence of the Semantic Conception of Evolutionary Theory. *Biology and Philosophy* 2: 26–32.

Thompson, P. 1988. *The Structure of Biological Theories.* Albany, NY: State University of New York.

Toulmin, S. 1974. The Structure of Scientific Theories. In: *The Structure of Scientific Theories*, ed. F. R. Suppe, 600–614. Urbana, IL: University of Illinois Press.

van Fraassen, B. C. 1970. On the Extension of Beth's Semantics of Physical Theories. *Philosophy of Science* 37: 325–334.

van Fraassen, B. C. 1972. A Formal Approach to the Philosophy of Science. In: *Paradigms and Paradoxes: The Challenge of the Quantum Domain*, ed. R. Colodny, 303–366. Pittsburgh: University of Pittsburgh Press.

van Fraassen, B. C. 1973. Semantic Analysis of Quantum Logic. In: *Contemporary Research in the Foundations and Philosophy of Quantum Theory*, ed. C. A. Hooker, 80–113. Dordrecht: Reidel.

van Fraassen, B. C. 1980. *The Scientific Image.* Oxford and New York: Oxford University Press.

van Fraassen, B. C. 1985. Empiricism and the Philosophy of Science. In: Churchland and Hooker (1985), 245–308.

van Fraassen, B. C. 1997a. Modal Interpretation of Repeated Measurement: Reply to Leeds and Healey. *Philosophy of Science* 64: 669–676.

van Fraassen, B. C. 1997b. Putnam's Paradox: Metaphysical Realism Revamped and Evaded. *Philosophical Perspectives* 11: 17–42.

van Fraassen, B. C. 1998. Review: J. Bub, *Interpreting the Quantum World. Foundations of Physics* 28: 683–689.

van Fraassen, B. C. 2008. *Scientific Representation: Paradoxes of Perspective.* Oxford: Oxford University Press.

van Fraassen, B. C. 2010. Rovelli's World. *Foundations of Physics* 40: 390–418.

van Fraassen, B. C. 2014. One or Two Gentle Remarks about Halvorson's Critique of the Semantic View. *Philosophy of Science* 81, 276–283.

van Fraassen, B. C. and Sigman, J. 1993. Interpretation in Science and in the Arts. In: *Realism and Representation*, ed. G. Levine, 73–99. Madison: University of Wisconsin Press.

von Neumann, J. 1931. Die Eindeutigkeit der Schrödingerschen Operatoren. *Mathematische Annalen* 104: 570–578.

von Neumann, J. 1955 [1932]. *Mathematical Foundations of Quantum Mechanics.* Princeton: Princeton University Press.

Weyl, H. 1940. The Ghost of Modality. In: *Philosophical Essays in Memory of Edmund Husserl*, ed. M. Farber, 278–303. Cambridge, MA: Harvard University Press.

Wheeler, J. A. 1957. Assessment of Everett's 'Relative State' Formulation of Quantum Theory. *Reviews of Modern Physics* 29: 463–465.

Steven French

Representation and Realism: On Being a Structuralist All the Way (Up and) Down

Abstract: Van Fraassen's *Scientific Representation: Paradoxes of Perspective* presents a structural empiricist approach to scientific representation. As he notes, in order to pin that representation down to an actual situation, we must supplement our "God-like" reflections on the structure with some further specifying feature. In this essay I consider what that further feature might be and argue that the realist has the resources to meet the challenge set by van Fraassen in the form of the "Appearances from Reality" criterion in the context of quantum mechanics. The issue then is whether such resources should be taken to be a constitutive part of that theory or not. Further reflection reveals that underpinning this whole discussion is a certain view of theories as things that have well-defined identity conditions and in conclusion I suggest that dropping this demand will be a liberating move in all sorts of respects.

> Do you desire to construct a vast and lofty fabric? Think first about the foundations of humility. The higher your structure is to be, the deeper must be its foundation. (St. Augustine, quoted after Gilbert 1895, 330)

1 Introduction: The Semantic Approach

It is a truism to say that yesterday's radical stance is today's establishment position and many have said this about the "Semantic Approach" to theories. Echoing Suppe from over 20 years before (Suppe 1989), Halvorson wrote that, "[w]ithin a few short decades, the Semantic Approach has established itself as the new orthodoxy" (Halvorson 2012; see also Frigg 2006; LeBihan 2012). The core claim of the Approach is taken to be that "a theory *is* [my emphasis] a class of models" (Halvorson 2012, 190; see also 192). On the basis of this claim Halvorson then goes on to present examples of what he calls theories that cannot be regarded in terms of classes of models and also models that should not be regarded as theories. As a result, he concludes that, "it is impossible to formulate good identity criteria for theories when they are considered as classes of models" (Halvorson 2012, 190; see also 201; for a response see Glymour 2013) and hence the Semantic Approach founders.

The nature of the above claim that a theory is a class of models and the issue whether it should be taken as characterizing the core of the Semantic Approach have of course been contested for many years. Thus, 30 years ago, van Fraassen

https://doi.org/10.1515/9783111019802-007

wrote, "if the theory as such, is to be identified with anything at all—if theories are to be reified—then a theory should be identified with its class of models" (van Fraassen 1989, 222). But of course, that's a big "if"! And one can, indeed some would say, *should*, resist this reification and instead take the Semantic Approach as just one such approach that can provide an array of devices—in this case set-theoretic models—that philosophers can deploy to capture certain aspects of scientific practice (see da Costa and French 2003; French and Saatsi 2006; French 2020b).

Furthermore, a crucial feature of that practice concerns the *representational* role of theories. As van Fraassen also notes, when a scientist presents a theory "she provides a class of models for the representation of those phenomena" (van Fraassen 2014, 277). He continues, "we properly speak of a model of combustion or of the San Francisco Bay in the way we speak of a painting of fire or of the Giaconda" (van Fraassen 2014, 277). Thus, "[a] model is a mathematical structure [of the kind considered by Halvorson] in the same sense that the *Mona Lisa* is a painted piece of wood" (van Fraassen 2014, 277). In other words, given that a scientific model is a representation, "it does not follow that the identity of a theory can be defined in terms of the corresponding set of mathematical structures without reference to their representational function" (van Fraassen 2014, 278). As French and Saatsi noted, "[i]t seems to be a popular misconception of the semantic view that it says *nothing* but the following about theories: theories *are* (with "is" of identity) just structures (models)" (French and Saatsi 2006, 552).

Granted that this is a misconception, noting that there is more to a theory than the mathematical structure immediately raises the question: how are we to capture this representational function *for our purposes as philosophers of science?* Again we can draw on the resources of the Semantic Approach and characterize it formally in terms of (partial) isomorphisms holding between the relevant structures (van Fraassen and Sigman 1993; van Fraassen 2008). Such a move has been criticized on the grounds that such devices are neither necessary nor sufficient for representation (see Frigg 2006; Suárez 2010) but these concerns can be met (Bueno and French 2011). That they are not necessary is argued for via examples, typically from the world of art, where, it is claimed, we have representation but no isomorphisms or structural similarity more generally. Leaving aside the relevance of such purported counter-examples when it comes to representation in science, Chakravartty (2001) has pointed out that it cannot be the case that similarity *in general* is not necessary for scientific representation, since in the absence of any similarities at all, it is not clear how the usual practices of interpretation and inference could get off the ground to begin with.

As for it not being sufficient, the argument is basically that representation is asymmetric—so da Vinci's painting represents Lisa Ghirardini but not vice versa—

whereas (partial) isomorphisms are not; again, responses can be sought, whether through appeals to inherent features of the formal relation or to "external" factors such as the intentions of the user of the representation (Bueno and French 2011). Here my focus is on the latter move, situated in a broader context. Consider the further question: does an appeal to intentions in such cases fatally undermine the Semantic Approach or a structuralist account of representation more generally? One might think that it only does so if one insists that *all* aspects of the relationship between theories and the world have to be understood structurally. However, consider theory confirmation: to demand that a structuralist account be given of *that* seems to ask too much.

This is the issue that I want to consider here, understood in broader terms. What formal devices such as partial structures and partial isomorphisms and the like provide us with is the framework for a structuralist account of scientific representation. The question now is, are these structures enough in general? Many have argued that they are not, as I have already indicated. In a sense I shall agree but I want to insist that the introduction of some further element does not undermine this sort of structuralist approach overall.

2 The Act of Self-Location

In his classic work on scientific representation van Fraassen adopted what has been called his "hauptsatz", namely that "[t]here is no representation except in the sense that some things are used, made or taken, to represent some things as thus or so" (2008, 23). Here we see such a non-structural element being introduced, namely the *user* of the representation. Is this fatal to the structuralist project of capturing, in the above terms, scientific theories and their representational function? One might balk at the claim that it is. After all, the structure, as presented above in terms of mathematical structures that are by their very nature abstract and general, must be effectively pinned down so that it can be said to apply to the particular situation under consideration. As van Fraassen himself puts it, "[t]he user must leave the God-like reflections on the structure [...] behind in order to apply the implications of those reflections to his or her actual situation" (2008, 71). This application, he suggests, must involve an essential act of self-location akin to the indicating element of a map that says "You are here".

This comparison with maps and models has been made before, of course, and one must be careful not to make too much of it. It is (typically) not the case that a scientist is simply given a theory in the way in which an orienteer, for example, is handed a map and told to find her way to some location, which of course requires her to know from where she is starting. Leaving aside the Kuhnian point about the education of scientists, they don't approach theories as one does a map, as a tool via which one can orient oneself in an unfamiliar landscape. A theory is an element or feature of the relevant scientific practice—it is part of the environment in which the scientist lives. Asking her where she stands with regard to the theory will surely result in the same bemused expression as obtained when asking someone where they are within their home neighbourhood!

Of course, it might be pressed that this is just to push the argument back a stage and that at some point some representation has to be produced towards which the scientist must orient herself. This could lead us into deeper waters than we might be comfortable paddling in here so let us keep our attention on theoretical models, such as we find in Einstein's Special Relativity, for example. As van Fraassen points out, if we are to use this theory to predict what happens to an electrically charged body in motion, say, then some choice of coordinate system related to a physical frame of reference is required. Leaving the God-like reflections on the structure of space-time behind in order to apply those reflections to the actual situation, or, in other words, for the representation to be *used*, something must be done that pins down the relevant structure to that situation. The connection between this choice of reference frame and the orientation of the scientist touched on above can be found in Weyl's famous characterization of coordinate systems as "the unavoidable residuum of the ego's annihilation" (quoted by van Fraassen in 2008, 71; for discussion of the phenomenological underpinnings of this phrase, see Ryckman 2005).

As van Fraassen goes on to note, such acts of self-location provide the conditions for relating theoretical models to specific empirical situations in general (2008, 87). And as we'll see, this becomes particularly pertinent in the case of that second great revolution in physics in the twentieth century, namely Quantum Mechanics and the infamous "measurement problem". Before we get there, however, let's consider measurement more generally.

3 Measurement as Representation

Consider yet another question: what sorts of interactions in general count as *measurements?* Clearly the outcome of a measurement must represent the target in a certain fashion, in the sense of selectively resembling it at a certain level of abstraction, according to the theory. This constitutes the representation criterion and, again, as van Fraassen puts it, "[t]he outcome provides a representation *of* the measured item, but also represents it *as* thus or so" (van Fraassen 2008, 180). An illustrative example (for my purposes at least) is provided by the images of collision events in particle detectors, such as those that yielded evidence for the existence of the Higgs boson at CERN, for example (see Garutti 2012). Here we might see a flash on the particle detector screen or computer generated lines whereby the particle under investigation is represented *as* an object and is granted, by virtue of being spatio-temporally located, at least for that instant, what Toraldo di Francia called a kind of “pseudo-individuality” (Toraldo di Francia 1978).

Why “pseudo-”? Well, when we try to project that particular representation beyond the measurement situation, as embodied in the particle detectors, we effectively bump up against quantum mechanics and must metaphysically accommodate, in particular, the implications of Fermi-Dirac and Bose-Einstein statistics (other kinds—the so-called parastatistics—are also theoretically possible and we’ll come back to those shortly). As is well-known, the likes of Born and Heisenberg (and others) took these statistics and the way that permutations of the particles appeared to be treated within the formalism as implying that quantum particles must be regarded as *non-individuals* in some sense (for details, both historical and formal, see French and Krause 2006). Cassirer and Eddington then took this, in turn, to imply that the particles could not be treated as *objects* at all and should be regarded as mere “nodes” in the relevant structure (Cassirer 1936; Eddington 1928). However, van Fraassen, Huggett, Redhead and French all argued that these quantum statistics are in fact also compatible with regarding the particles as individuals (by virtue of a different interpretation of particle permutations; see French and Redhead 1988; Huggett 1999; van Fraassen 1989). With Krause arguing that non-individuality can be formally characterized via quasi-set theory without abandoning the view of the particles as objects, that connection with structuralism was broken. In its place, Ladyman argued that it is the metaphysical *underdetermination* that results from these developments—between particles-as-individuals and particles-as-non-individuals—that represents a challenge for realism (see also van Fraassen 1989) and that this challenge could be met, and the underdetermination broken, by shifting from an object-oriented metaphysics to “ontic structural realism” (Ladyman 1998; French 2014).

4 Realism (of a Structural Kind)

In his critique mentioned above, Halvorson claims that structural realism actually derives its plausibility from the Semantic Approach and suggests that, given his criticism of the latter, “structural realism is no more plausible today than it was in the time of Carnap and Hempel” (Halvorson 2013, 476). However, although Ladyman (1998) did acknowledge that the Semantic Approach might be preferred by the structural realist because, in effect, it wears its structuralist colours on its sleeve, Worrall’s so-called “epistemic” form of structural realism was explicitly couched in terms of the alternative "Syntactic Approach" that characterizes theories in terms of sets of propositions, closed under some (typically classical) logic (Worrall 1989). Furthermore, Bain and Landry have argued that category theory is in fact a better framework for the structural realist to adopt (Bain 2013; Landry 2007; for criticisms see Lam and Wuthrich 2015) and French has acknowledged that the structuralist has a variety of meta-representational options to choose from (see French 2014, Ch. 5). Thus the ties with the Semantic Approach are looser than Halvorson suggests!

Furthermore, historically and currently, there were multiple motivations for structural realism, from which it can be said to derive its plausibility. For Worrall (1989) and Saunders (1993) the motivation was accommodating theory change and responding to the challenge set by Laudan, for example. For Cassirer and Eddington, the motivation for their neo-Kantian and "subjective" structuralisms, respectively, had more to do with accommodating what theories of modern physics, such as General Relativity and Quantum Mechanics, appear to tell us. Similarly for French (2014) and Ladyman (1998), there is a significant emphasis on accommodating the role that certain symmetry principles play in today's particle physics, including, of course, the Permutation Symmetry that lies behind the Bose-Einstein and Fermi-Dirac statistics mentioned above. For them, the significance of this role tells us something important about the *world*, whereas for van Fraassen, of course, as a structural empiricist, it tells us something about *science*.

And what it tells us, I would maintain, is that the world, conceived *as* structure, should be understood, following Cassirer, as akin to a Parmenidean "well-rounded" sphere in which symmetries, laws and, crucially, measurement outcomes are all interlocked (French 2014). The contents of those (possible) measurement outcomes constitute what van Fraassen calls the "appearances" and here again we return to the overall theme of this essay: these outcomes function as what Wilson calls the "existential witnesses" (Wilson 2017) that "pin down" the structure of the world, as indicated above, to that which is applicable to our situation. So, consider quantum statistics yet again: Permutation Symmetry yields a range of possible statistical kinds, as again already indicated, including not just Bose-Einstein and Fermi-Dirac statistics but also an infinite variety of parastatistics (the application of these to particle physics played a crucial role in the development of quantum chromodynamics). However, we appear to occupy only the bosonic and fermionic sectors of Hilbert space and not any of the parastatistics ones, so the structure of the world is one that manifests bosonic and fermionic behaviour and in that sense is "pinned down" to those sectors.

Recalling what was touched on above about indexicality and what van Fraassen calls the Act of Self-Location, it appears that here too we have the insertion of an apparently non-structural feature into our structuralist picture that effectively says "we are *here!*" (in Hilbert space). Does that self-locating move undermine the structuralist enterprise? No—that aspect of the structure of the world that incorporates Permutation Symmetry allows for a range of possibilities and in order to talk about the structure of *this* world we need to specify which of those possibilities is actualized. As in the case of theory confirmation, the specification of the actual world from an array of possibilities should not itself be up for structuralist interpretation.

Acknowledging that we have more structural resources on hand than we need to represent the actual world also has obvious implications for discussions of modality in science. For the structural realist, the world, as structure, can be regarded as inherently modal, offering, perhaps, an alternative to the Humean and dispositionalist accounts of laws and symmetries (see Berenstain and Ladyman 2012; French 2014). For the empiricist, any modality is to be understood as "in" the models; that is, as simply a reflection of the fact that, as just stated, the mathematical characterization of Permutation Symmetry allows for a number of possibilities. In either case, acknowledging this feature allows us to further extend our explorations of the resources that are made available in the context of modern physics.

Nevertheless, structural realism faces a further serious challenge, which again illuminates the theme of how we "pin down" our structural commitments and is presented by van Fraassen as follows.

5 Challenge: The Appearance from Reality Criterion

According to van Fraassen, the realist must be committed to the claim that the scientific representation of nature should include the appearances by showing how they are "produced as a proper part of the depicted reality" (van Fraassen 2008, 281). Here he quotes Leplin:

> A theory is not simply an empirical law or generalization to the effect that certain observable phenomena occur, but an explanation of their occurrence that provides some mechanism to produce them, or some deeper principles to which their production is reducible. (Leplin 1997, 15)

Thus what is required is an *explanation* that takes us from the relevant feature of the theory, as *explanans*, to the appearances, as *explanandum*, that allows us to grasp how the latter is "produced" by the former; or as van Fraassen puts it, "[t]he appearances are to be explained as *produced* in the world depicted by fundamental physics" (2008, 282). Here, then, is the challenge for the structural realist:

can the relevant symmetries, such as Permutation Symmetry, act as these "deeper principles" and function as the *explanans* in an explanation of the relevant appearances?

I think the answer to that question is "yes" and so the challenge can be met. Consider the following phenomenon: white dwarf stars are the final evolutionary stage of stars that are not sufficiently massive to form neutron stars and that stage is reached when the gravitational collapse of the star is halted by electron degeneracy "pressure". The explanation for that "halting" of the collapse appeals to the Pauli Exclusion Principle which prevents electrons from occupying the same state (where their state encompasses not only their energy but also their spin). How the Exclusion Principle, viewed as a constraint on electron occupancy, can act as an *explanans* in this case has been a matter of some dispute (Lewis 1986; Skow 2014). However, French and Saatsi (2018) have argued that we can accommodate this case within Woodward's "counterfactual-dependence" account of explanation, in which the core idea is that symmetries play a role in explanatory arguments that is comparable to that of a contingent initial or boundary condition in causal explanations. Thus, a symmetry fact together with an appropriate connection between that fact and the *explanandum* can provide "what-if-things-had-been-different" information, thereby showing how an explanandum *depends on* the symmetry. Although this counterfactual-dependence view of explanation has been largely developed in the context of causal dependence, in recent years it has been extended to various kinds of non-causal dependencies (see, for example, Jansson and Saatsi 2019; Saatsi 2018) and many symmetry explanations are likewise naturally construed as being non-causal.

Of course, there remains the issue of the kind of dependence involved if it is non-causal and this bears on van Fraassen's insistence that the appearances must be explained as *produced* in the world—clearly when it comes to symmetries the mode of production must go beyond the causal or "mechanistic" in Leplin's terms. Here we must step carefully so as to not beg any questions, particularly against the structural realist! First of all, we must appropriately characterize the relationship between Permutation Symmetry and Fermi-Dirac statistics, the nature of which is captured by the Pauli Exclusion Principle that features in the above explanation of the halting of white dwarf collapse. There is a range of devices in the metaphysicians' "toolbox" (French and McKenzie 2012) that we might utilize to capture the nature of this relationship but some can be shown to be unfit for purpose in this particular case (McKenzie 2020; Wolff 2012). As an alternative, we might deploy the determinable-determinate relationship (Wilson 2017), with Fermi-Dirac statistics understood as the determinate of Permutation Symmetry, construed as the relevant determinable (French 2014). We can even formalize this relationship in such a way as to capture the idea of the determinate "belonging" to the determinable (Denby 2001).

This then captures the sense in which the relevant symmetry plays the role of a boundary condition as mentioned above—we exist in a world bounded by the Bose-Einstein and Fermi-Dirac sectors of Hilbert space. However, this still leaves the further question: how does the fact that electrons are fermions and subject to the Exclusion Principle *explain* the halting of the star's collapse? Skow, for example, takes the phrase "degeneracy pressure" seriously, arguing that we can extend the meaning of "pressure" from that of the impact of gas atoms on some container, to that of the resistance that results from electrons occupying the relevant quantum states (Skow 2014). However, given that the "action" of the Exclusion Principle cannot be accounted for in terms of any of the known fundamental forces, it is clear that such an extension effectively releases the meaning of the term from its causal moorings (French and Saatsi 2018, 192). Strevens, on the other hand, suggests that the relationship between the Exclusion Principle and the halting should be understood as some kind of metaphysical dependence, as in the case of Permutation Symmetry above (Strevens 2008, 178). However, in this case we do have the kind of asymmetry we associate with dependence, unlike the relationship between Permutation Symmetry and Fermi-Dirac statistics, in that the halting of the collapse of the white dwarf star is dependent on the action of the Exclusion Principle and not vice versa. Of course, you might find this too "thin" a characterization, as compared to a causal relationship but here a structural realist might suggest that this thinness is a result of the metaphysical proximity of the two relata: for her, the observed halting and the Exclusion Principle are both features of the structure of the world, conceived à la Cassirer as indicated above.

There is more to say, of course but I would maintain that Permutation Symmetry can be said to "produce" the appearances associated with the halting of the collapse of white dwarf stars in a way that meets van Fraassen's challenge. However, van Fraassen then sharpens the problem for the realist by posing that challenge in the context of quantum mechanics, where it appears in the form of the infamous "measurement problem".

6 The Measurement Problem

Thus he suggests that "the Copenhagen development of quantum theory exemplifies a clear rejection of the [Appearances from Reality] criterion" (van Fraassen 2008, 291). And given the tremendous success of the theory, under that interpretation, we should then be wary of imposing this criterion in general. If the realist nevertheless insists on accepting it, then she must deal with the measurement problem, which van Fraassen characterizes as follows:

> Measurement Problem: the quantum theoretical description of the interaction between system and measurement apparatus provides no place for specific outcomes (see van Fraassen 2008, 300).

Putting things quite bluntly, quantum mechanics offers an inherently probabilistic description (think of Schrödinger's Cat) and in order to account for a specific measurement outcome (such as "the cat is alive"!) something further must be added to the theory, where, by virtue of being extraneous to the theory, that "something further" goes beyond what can be supported by the physics and hence can be contested. I'll come back to that shortly but as van Fraassen notes, the measurement problem is not an issue for empiricists, since the empirical adequacy of the theory can be cashed out in terms of the "surface models of phenomena fit[ting] properly with or into the theoretical models" (2008, 305); that is, we have matching between two families of probability functions. From this perspective, he goes on, the measurement problem represents a "methodological rejection" of the Appearance from Reality criterion in the new theoretical context of quantum physics.

As he also emphasizes, it is likewise no problem for physicists who (more or less) happily continue with their theoretical and experimental practices without (for the most part) worrying about the problem. As Cordero has argued, they have decoherence to thank for that, which introduces an "effectively irreducible 'experimental astigmatism'" (Cordero 2001). "Decoherence" is a term that, very broadly speaking, describes the "loss" of quantum behaviour when a system interacts with the environment As a result,

> [a]spects as profound as those regarding the group-theoretic symmetries are untouched by the debate at the stochastic level. That is, above a certain descriptive depth all the models yielded by the [...] [various interpretational] approaches converge both structurally and semantically in terms of effective partial isomorphisms that reach deeply into the respective theoretical fabrics [...]. (Cordero 2001, 308)

In effect, then, different interpretations of the theory—such as the Bohm, Everett and GRW interpretations, to pick the more well-known—share a thick body of modelling and relevant prior knowledge in terms of which the outcomes produced by the sorts of particle detectors mentioned above, say, can be understood. Cordero argues that this yields a series of "local meanings" of the theory that can be captured via partial representations involving restricted domains that in turn incorporate limited levels of theoretical and experimental precision. What this then supports is, at best, a "restricted" form of structural realism that resonates with recent moves to "local" forms of realism (Saatsi 2017; Vickers 2013). These in turn urge us to limit our realist commitments to those that are specific to the relevant theoretical situation under consideration.

Such "localized" realism is not necessarily at odds with the structuralist tendency in general, especially if we recall the core motivation, running from Cassirer and Eddington to Ladyman and current accounts, of accommodating certain specific features of twentieth-century physics. Alternatively, however, we might still prefer a more global account and consider, in a little more detail, the nature of that "further element" that should be added to the quantum-mechanical description in order to satisfy van Fraassen's "Appearances from Reality" criterion. The hope then is that this will bridge the gap between the probabilistic description offered by the theory and the specific measurement outcomes, in some way or other. As we'll see, this can be done but only at some cost.

Let us begin with the Copenhagen development of the theory, as mentioned above. As is now well-known the label was actually applied in the 1950s (Beller 1999) as a way of characterizing what was deemed to be the "orthodoxy"; indeed, the labelling may be said to have contributed to the cementing into place of that orthodoxy. And that has since been presented in broadly instrumentalist terms that are amenable both to physics practice and empiricist philosophy. However, prior to that labelling moment, one can discern distinct strands within the groups of physicists most commonly associated with the so-called "Copenhagen Interpretation", offering two types of solution to the measurement problem and thus two ways of bridging the above gap: the first is associated with Bohr and appeals to the "classical" nature of the measurement environment. Again putting things simply, the core idea here is that we are compelled to describe measurement outcomes in classical terms and hence the "gap" between what the theory says and specific outcomes isn't bridged so much as entrenched in a fundamental shift from one kind of description to another. All we need to do, then, is make sure we use the right description in the right context. As is well-known, this solution came under intense criticism, not least on the grounds that as it continued to be developed, quantum mechanics was actually applied to broadly "macroscopic" situations for which "classical" physics was supposed to be appropriate, thus undermining the force of the "compulsion" to use classical language in such situations. A striking and, at the time, forceful example of this is the development, from the mid-1930s onwards, of quantum models of superconductivity and superfluidity, led by the London brothers (another, presented in the form of a thought experiment, is Schrödinger's Cat, as touched on earlier).

The second strand is typically associated with von Neumann and is generally understood as resolving the measurement problem by appealing to consciousness as the "extra factor". This bridges the gap by "collapsing" the quantum wave function, reducing the range of possible outcomes to just one. Well-known and obvious concerns were raised about this proposal, not least having to do with how the intervention of consciousness could "cause" the wave function to collapse or more generally close the gap between the quantum description and the specific measurement outcomes. However, von Neumann himself actually wrote very little on the role of consciousness and many of those concerns were expressed in response to what was taken to be a summary of his view, written by London and Bauer (1983 [1939]). As it turns out, this was no mere summary but presented an entirely different approach that drew on London's background in Husserl's phenomenology (and here we recall Ryckman's contextualization of Weyl's point, acknowledged by van Fraassen, about coordinate systems in the relativistic context). Instead of seeing consciousness as some extraneous element that mysteriously collapses the wave function and thereby accounts for specific measurement outcomes, it is here presented within an understanding of quantum mechanics as a "theory of knowledge" in general. Within that framework, the wave function is taken to describe *everything*, contra Bohr, and it is the separation of both consciousness and the system under consideration from the superposition through a phenomenological act that yields a definite measurement result. Hence what we have is not so much a further or extraneous element imposed from outwith the theory, as it were, but a phenomenologically grounded expression of the very conditions for doing science, indeed for acquiring knowledge, in the first place. Within this framework, the measurement problem is seen as no "problem" at all, but rather as an acute manifestation of the nature of our engagement with "the world" that can only be properly grasped in phenomenological terms (see French 2020a and French forthcoming). It was, in fact, Wigner who effectively appropriated this account and presented it in terms of consciousness mysteriously producing the collapse of the wave function, thereby generating all the concerns that ultimately led to the approach in general being widely discarded.

So, when it comes to the Copenhagen development of quantum theory it appears we have (at least) three variants: Bohr's version, with a classicality condition; von Neumann/Wigner's with consciousness as the cause of collapse; and London's with the theory understood from a phenomenological perspective. Only in the second case do we have a specific extraneous element that is introduced; in the other two we have, rather, the articulation of conditions for applying the theory or acquiring knowledge more generally. In whatever case, where do we stand with the Appearance from Reality criterion? Recall that the requirement is that the appearances should be explained as *produced* in the world depicted by the physics. Now, if the mode of production is taken to be mechanistic, then only in the case of the von Neumann/Wigner variant could we say that the requirement is met and then only contentiously and with a broad understanding of "mechanistic". With that meaning of "produce" on the table, it would indeed seem that when it comes to the cases of the Bohr and London variants, the Appearance from Reality criterion has been set aside. However, to leave it at that would be to ignore the role of the further conditions which are introduced in each case—a classical/quantum division in that of Bohr and a fundamentally phenomenological framework in the case of London. It is certainly not the case that "the theory", however that is to be understood (and we'll come back to this issue shortly), yields the appearances on its own. In either case, some further move has to be made and although we obviously can't give this a mechanistic construal, perhaps we can take some liberties and acknowledge such a move as a form of production, in broadly van Fraassen's sense. If so, then it would seem that the criterion can be met, and in an interestingly different way in each case. However, if we acknowledge this, there is the further question to be answered, as just touched upon: what, then, *is* the theory of quantum mechanics?

Recalling Halvorson's critique of the Semantic Approach that we began with, if we were to move away from the toy examples that he focuses on and characterize an actual theory in terms of families of models, such as quantum mechanics, what would we start with? Do we include the above "further elements" and conditions? In Bohr's case, presumably not, since his classicality condition can be understood as delineating the domain of applicability of "the" theory but in the London case, there is no such need for delineation; rather, it is only when we see "the" theory from the phenomenological perspective that we can account for the appearances. But then, what do we take "the" theory to be?

Everett, at least, was clear in his response to what was perceived to be the Copenhagen orthodoxy. Under the influence of Frank's "empirical pragmatism" (Barrett and Byrne 2012, 258) he maintained that a theory, in general, is a mathematical model with a homomorphism holding between the model and "the world", where the latter could either be taken to be "the world of experience" (i. e. the sense perceptions of the individual), or the "real world", " depending upon one's choice of epistemology" (Everett in his "Long Thesis"; Barrett and Byrne, 169). Here we have a characterization that obviously meshes with the Semantic Approach, but with homomorphism selected as the appropriate formal relation rather than isomorphism because Everett felt that not only was the theory not required to explain *all* of our experience but there may be parts of the model that are not interpreted *as* our experience (Barrett and Byrne 2012, 169, fn cq). Crucially for the interpretation of quantum mechanics that he proposed, what he meant by "experience" here were memory sequences and more particularly, those that were "typical" (in a sense that when technically defined gave the standard quantum statistics; see Barrett 2018).

As a result, Everett insisted, despite the kinds of concerns that underlie the challenge represented by the "Appearance from Reality" condition, "[t]he theory *is* isomorphic with experience when one takes the trouble to see what the theory itself says our experience will be" (Everett to DeWitt, in Barrett and Byrne 2012, 255). Thus, he maintained, what he called "pure wave mechanics", with its superpositions of states, is in fact "empirically faithful" as long as one understood that there has to be a "renegotiation" of empirical adequacy in the context of theory selection. Now of course, this notion of faithfulness offers only a weak empirical standard as Barrett has noted and there is no explanation of why a particular sequence of records should be regarded as "typical" (Barrett 2018). Nevertheless, Everett's account does show how the appearances might be obtained from the fundamental physics. It also, crucially, highlights how the satisfaction of the Appearances from Reality criterion assumes implicit prior consideration of both what we take to be "the" theory and how we conceive of its relationship to experience.

Pressing on in our search for a global realist response to van Fraassen's challenge, we can re-frame this account in broadly structuralist terms. The central premise of (ontic) structural realism aligns with Everett's overall stance, namely to take quantum mechanics seriously as presenting "the structure of the world". For the Everettian this structure has to be understood in terms of "branches" within the framework of the relative-state formulation but in a sense it offers too much structure. Again, some further feature or device has to be introduced to pin down the relevant aspects of that structure to yield our experiences. In Everett's case this feature is the relevant memory sequences, with measurement outcomes "typically" distributed according to the standard quantum probabilities. Thus this amounts to "what the theory itself says our experience will be" (Everett to DeWitt, Barrett and Byrne 2012, 255). As noted above, the associated empirical faithfulness may seem too modest, explanatorily speaking. In that case, one might prefer the so-called Deutsch-Saunders-Wallace approach which introduces decision theory as the further element, yielding the relevant probabilities via reflection on the choices of an ideal agent (Wallace 2012).

Again, we can usefully compare such "further elements" to those indicated above when it comes to the different variants of the Copenhagen Interpretation. The lesson to be drawn is that given the nature of what Everett called the "mathematical model" sitting at the heart of "the" theory, something further has to be added to the mix, whether that's consciousness, or typicality or, thinking of the Bohmian case, some privileged "hidden variable" such as position or some kind of probability "field" as with GRW, or a reconceptualization of quantum mechanics as itself a theory of knowledge, as London and Bauer maintained. All of these features or moves can be understood as enabling the "production", perhaps understood broadly, of the appearances from the physics. And crucially, then, the challenge underlying the "Appearance from Reality" criterion hinges on a prior decision as to what we are going to take "the" theory to be. That decision in turn hinges on "what explanatory assumptions one has an obligation to include in one's specification of a theory" (Barrett 2017, 38). In particular, if one feels that one is not obliged to include any of the above elements, moves or frameworks in one's specification of quantum theory, then "the" theory by itself, as it were, cannot meet van Fraassen's requirement. If, on the other hand, one takes these features precisely to be the kinds of explanatory assumptions that should be part of the specification, then the challenge can be met, but with obvious costs, not only for each case but also globally, with the blossoming of a multiplicity of different quantum "theories". And lest you think that this is just underdetermination redux, compare and contrast Everett's "pure wave mechanics" plus typicality with London's phenomenologically based conception of quantum mechanics as a theory of knowledge —can these usefully be regarded as merely different interpretations?

It seems, then, that we have something of a dilemma: either we exclude such explanatory assumptions from our specification of "the" theory, and then bite the empiricist bullet when it comes to obtaining the appearances from reality, or we introduce some such assumption and accept both the local and global costs. In seeking a way forward, let us return to Halvorson's critique of the Semantic Approach. We recall that van Fraassen's response was to emphasize the significance of the representational function of theories. Here the example of the Mona Lisa was invoked and, as we noted above, scientific theories are typically compared with paintings in discussions about representation, not always appositely. And indeed, more generally, theories are not like paintings in certain, obvious, ways— they are not physical objects, for example. This is clearly fundamental when it comes to the evaluation of the faithfulness of the representation in each case. Furthermore, it is not just that a theory cannot be identified with a particular presentation of it, in a particular "mode" or language, say. We would not want to say that Newtonian mechanics expressed in French is a different theory from that expressed in English. Likewise, many would be reluctant to claim that the Special Theory of Relativity as expressed in Einstein's 1905 paper in terms of clocks and rods is a different theory from the formulation presented in terms of Minkowski space-time, although that might be more contentious than in the Newtonian case. Indeed, it is these kinds of "intuitions" that motivate, in part, the Semantic Approach—*if* theories are to be identified with anything at all (and again, that's a big "if"), they should be identified with something that transcends language.

Alternatively, theories have been compared with other forms of art, such as works of music. Popper, famously, placed both theories and musical works in his "World Three", the world of "intelligibles"; that is, the products of the human mind (whereas World One is the world of physical states and World Two is that of mental states). This encourages the view that they are in some sense abstract entities, or, perhaps, abstract *artefacts* as Thomasson has suggested, created and sustained by certain intentions (Thomasson 2006). The worry here is that not only do such views offer an inflated ontology but they face a number of challenges in terms of accommodating the kinds of heuristic moves that can be identified and traced in scientific practice (see French 2020b).

There is a further alternative to all these sorts of accounts that avoids having to introduce new worlds, or abstract artefacts or, indeed, any further ontological commitments at all. It also allows us to escape the above dilemma and effectively sidesteps the need to proscribe identity conditions for theories that the likes of Halvorson demand. We simply deny that theories are the sorts of "things" that have identity conditions at all, whether abstract artefacts, denizens of World Three or whatever. Here we also recall van Fraassen's quote from before: "if theories are to be reified [...] then a theory should be identified with its class of models" (van Fraassen 1989, 222). As I've said, that's a big "if" and what is being suggested here is that we bluntly decline to affirm the antecedent. One way of making good on this is simply to *deny* that there "are", in whatever sense, theories at all (French and Vickers 2011). Of course we may say things that are apparently *about* theories, such as "quantum mechanics is our most empirically successful theory" or "quantum mechanics is elegant" but that should not be taken to imply that what makes these statements true is some *thing*, "the" theory of quantum mechanics. We can adopt the eliminativist strategy of taking something else as the truthmakers of such statements, such as the various practices engaged in by physicists (and philosophers, as we'll see; French and Vickers 2011; French 2020b).

So what makes the statement "quantum mechanics is our most empirically successful theory" true are a range of practices associated with experimental work, as well as with modelling and theoretical manipulation. Of course, in articulating that sense of empirical success we will have to address the kinds of issues that Everett tried to tackle and that underpin the Appearance from Reality criterion. So we should include among those practices the broadly reflective practices of the likes of Everett, as well as London, von Neumann and others, that involve thinking about selecting "typical" memory sequences, or the insertion of consciousness into our framework or even taking "the" theory to be a theory of knowledge under the phenomenological understanding. As for "quantum mechanics is elegant", we can cash this out in terms of some combination of parsimony and power and take the statement to be made true by the relevant practices, involving, for example, the ease of deduction of certain statements from the axioms or fundamental claims of the theory, the way in which a wide variety of claims (both theoretical and empirical) can be obtained from these axioms and so on (French 2020b).

This eliminativist stance avoids a bloated ontology but potentially at the cost of rendering problematic our understanding of theories as *representations*. Now we face yet another question: how can something that is not a *thing* represent something else?! The answer is to acknowledge that when we, whether scientists or philosophers of science, make claims about theories fulfilling a representational role we have a certain construct in mind. We may then present that construct to one another, via symbols scribbled on paper or the whiteboard, or, particularly if we are philosophers, via some framework such as that offered by the Semantic Approach. However, the models of the latter are neither to be identified with the theory nor to be taken as representing that theory as an entity, whether abstract or not; rather, they should be seen as devices that we, philosophers of science, can deploy to (meta-)represent those aspects of scientific practice we are interested in. At that (meta-)level we can then introduce partial isomorphisms and the like to capture the relevant representational function holding between "the theory" (meta-characterized in terms of a family of models) and the given system (also meta-characterized in terms that allow for such a relation to be taken to hold; see French 2020b).

Expressing it in those terms raises an obvious worry for the realist, but I want to emphasize that to deny the existence of theories as things is not to deny the existence of things apparently referred to by those theories. We should not allow our eliminativism to bleed across from our stance on theories to our stance towards the world. Nevertheless, we might well have good reason to be a certain kind of eliminativist in the latter case also. Consider: certain scientific practices running throughout twentieth-century physics emphasize and illuminate the role of symmetries and, as indicated above, this has been one of the prime motivating forces behind structuralism, from Cassirer and Eddington to Ladyman. Viewed, realistically, as saying something about the world, we can take structuralism and, in particular, the fundamental role granted to symmetries, as implying the elimination of elementary particles *as things*, with statements apparently about them made true by features of the structure of the world (French 2014).

However, that's a whole other set of issues and one can (and should) be an eliminativist about theories while adopting any of a range of stances and positions within those stances towards the ontology of the world. Having said that, it should now be apparent that there is a certain parallel between the issue of "pinning down" the structure of the actual world and that of satisfying the Appearance from Reality criterion. With regard to the former, we noted that the mathematical description of these symmetries yields surplus structure that goes beyond what we need in practice. So, again, the group-theoretic description of Permutation Symmetry yields forms of quantum statistics that do not seem to be relevant to our world. Thus we need to pin down that structure to the actuality and understand Bose-Einstein and Fermi-Dirac statistics as "existential witnesses", in Wilson's sense. Likewise, as we've just considered, when it comes to quantum mechanics we have, in a sense, too much structure and we need to introduce some further element or device to yield the appearances, whether that be "typicality", the precepts of decision theory, consciousness or whatever. In both cases, it seems, we have to engage in a kind of "Act of Self-Location" as a form of making concrete the abstract structural forms that the mathematics offers us.

7 Final Remarks

Of course, these moves of "pinning down" the structure and establishing "existential witnesses" are fallible and revisable across a range of respects and degrees. Thus, nothing said here goes against van Fraassen's insistence that "we have the freedom to follow the contemporary abstract structural forms now prevalent in the advanced sciences" (2008, 267). Indeed, exercising that freedom is a source of significant scientific development. Consider Dirac's famous equation in relativistic quantum mechanics: putting things very simply, from a mathematics perspective this has both positive and negative energy solutions. The latter might be understood as constituting "surplus" structure that can be dismissed as non-physical, thereby "pinning down" the physical solutions. However, as is well known, and again summarizing things rather crudely (for more details see Bueno and French 2018, Ch. 7), Dirac exercised the freedom that van Fraassen mentions and interpreted this particular surplus structure in terms of what we now understand as anti-matter. As Redhead emphasized many years ago, we can find this exploration and interpretation of surplus structure time and again in the history of physics (Redhead 1975). Sometimes it pays off, as in Dirac's case and sometimes it doesn't, as when quarks were briefly characterized as paraparticles. In both cases, however, we can see the "pin" sliding, as it were, across the structures that are available. Indeed, we can perhaps see that history as a series of such explorations together with appropriate "pinning down" to yield descriptions of either—depending on your stance—how the world is or how it could be. And, I would insist, what we should focus on, when it comes to the philosophy of science, are the *practices* associated with these moves, and, returning to the opening remarks of this essay, what we should acknowledge is the diversity of devices that we, as philosophers of science, can deploy to make sense of these practices.

References

Bain, J. 2013. Category-Theoretic Structure and Radical Ontic Structural Realism. *Synthese* 190: 1621–1635.

Barrett, J. A. 2017. Typical Worlds. *Studies in History and Philosophy of Science Part B: Studies in History and Philosophy of Modern Physics* 58: 31–40.

Barrett, J. A. 2018. Everett's Relative-State Formulation of Quantum Mechanics. *The Stanford Encyclopedia of Philosophy* (Winter 2018 Edition), ed. E. N. Zalta. https://plato.stanford.edu/archives/win2018/entries/qm-everett (last accessed: 30. 5. 2023).

Barrett, J. A. and Byrne, P. (eds.). 2012. *The Everett Interpretation of Quantum Mechanics: Collected Works 1955–1980 with Commentary.* Princeton: Princeton University Press.

Beller, M. 1999. *Quantum Dialogue: The Making of a Revolution.* Chicago: University of Chicago Press.

Berenstain, N. and Ladyman, J. 2012. Ontic Structural Realism and Modality. In: *Structural Realism*, eds. E. Landry and D. Rickles, 149–168. Dordrecht: Springer.

Bueno, O. and French, S. 2011. How Theories Represent. *The British Journal for the Philosophy of Science* 62: 857–894.

Bueno, O. and French, S. 2018. *Applying Mathematics: Immersion, Inference, Interpretation.* Oxford: Oxford University Press.

Cassirer, E. 1936. *Determinismus und Indeterminismus in der modernen Physik.* Göteborg: Wettergren & Kerber. Trans. as *Determinism and Indeterminism in Modern Physics.* New Haven: Yale University Press, 1956.

Chakravartty, A. 2001. The Semantic or Model-Theoretic View of Theories and Scientific Realism. *Synthese* 127: 325–345.

Cordero, A. 2001. Realism and Underdetermination: Some Clues from the Practices-Up. *Philosophy of Science*, S68: 301–312.

Da Costa, N. C. A. and French, S. 2003. *Science and Partial Truth.* Oxford: Oxford University Press.

Denby, D. A. 2001. Determinable Nominalism. *Philosophical Studies* 102: 297–327.

Eddington, A. 1928. *The Nature of the Physical World.* Cambridge: Cambridge University Press.

French, S. 2014. *The Structure of the World: Metaphysics and Representation.* Oxford: Oxford University Press.

French, S. 2020a. From a Lost History to a New Future: Is a Phenomenological Approach to Quantum Physics Viable? *Phenomenology and Physics*, 205–225, eds. H. Wiltsche and P. Berghofer, Cham: Springer.

French, S. 2020b. *There Are No Such Things as Theories.* Oxford: Oxford University Press.

French, S. (forthcoming). *A Phenomenological Approach to Quantum Mechanics: Cutting the Chain of Correlations.* Oxford: Oxford University Press.

French, S. and Krause, D. 2006. *Identity in Physics.* Oxford: Oxford University Press.

French, S. and McKenzie, K. 2012. Thinking Outside the (Tool)Box: Towards a More Productive Engagement between Metaphysics and Philosophy of Physics. *The European Journal of Analytic Philosophy* 8: 42–59.

French, S. and Redhead, M. 1988. Quantum Physics and the Identity of Indiscernibles. *British Journal for the Philosophy of Science* 39: 233–246.

French, S. and Saatsi, J. 2006. Realism about Structure: The Semantic View and Non-linguistic Representations. *Philosophy of Science (Proceedings)* 78: 548–559.

French, S. and Saatsi, J. 2018. Symmetries and Explanatory Dependencies in Physics. In: *Explanation Beyond Causation: Philosophical Perspectives on Non-causal Explanations*, eds. A. Reutlinger and J. Saatsi, 185–205. Oxford: Oxford University Press.

French, S. and Vickers, P. 2011. Are There No Such Things as Theories? *The British Journal for the Philosophy of Science* 62: 771–804.

Frigg, R. 2006. Scientific Representation and the Semantic View of Theories. *Theoria* 21: 49–65.

Garutti, E. 2012. The Physics of Particle Detectors. Lecture Notes. https://www.desy.de/~garutti/LECTURES/ParticleDetectorSS12/L1_Introduction_HEPdetectors.pdf (last accessed: 26.6.2023).

Gilbert, J. H. 1895. *Dictionary of Burning Words of Brilliant Writers.* W. B. Ketcham: New York.

Glymour, C. 2013. Theoretical Equivalence and the Semantic View of Theories. *Philosophy of Science* 80: 286–297.

Halvorson, H. 2012. What Scientific Theories Could Not Be. *Philosophy of Science* 79: 183–206.

Halvorson, H. 2013. The Semantic View, If Plausible, Is Syntactic. *Philosophy of Science* 80: 475–478.

Huggett, N. 1999. Atomic Metaphysics. *Journal of Philosophy* 96: 5–24.

Jansson, L. and Saatsi, J. 2019. Explanatory Abstractions. *British Journal for the Philosophy of Science* 70: 817–844.

Ladyman, J. 1998. What Is Structural Realism? *Studies in History and Philosophy of Science* 29: 409–424.

Lam, L. and Wuthrich, C. 2015. No Categorial Support for Radical Ontic Structural Realism. *British Journal for the Philosophy of Science* 66: 605–634.

Landry, E. 2007. Shared Structure Need Not Be Shared Set-Structure. *Synthese* 158: 1–17.

LeBihan, S. 2012. Defending the Semantic View: What It Takes. *European Journal for the Philosophy of Science* 2: 249–274.

Leplin, J. 1997. *A Novel Defence of Scientific Realism.* Oxford: Oxford University Press.

Lewis, D. K. 1986. Causal Explanation. In: *Philosophical Papers*, Vol. 2, 214–240. Oxford: Oxford University Press.

London, F. W. and Bauer, E. 1983 [1939]. The Theory of Observation in Quantum Mechanics. In: *Quantum Theory and Measurement*, eds. A. Wheeler and W. H. Zurek, 217–259. Princeton: Princeton University Press.

McKenzie, K. 2020. Structuralism in the Idiom of Determination. *The British Journal for the Philosophy of Science* 71, 497–522.

Redhead, M. 1975. Symmetry in Intertheory Relations. *Synthese* 32: 77–112.

Ryckman, T. 2005. *The Reign of Relativity.* Oxford: Oxford University Press.

Saatsi, J. 2017. Replacing Recipe Realism. *Synthese* 194: 3233–3244.

Saatsi, J. 2018. On Explanations from "Geometry of Motion". *British Journal for the Philosophy of Science.* 69: 253–273.

Saunders, S. 1993. To What Physics Corresponds. In: *Correspondence, Invariance and Heuristics: Essays in Honour of Heinz Post*, eds. S. French and H. Kamminga, 295–326. Dordrecht: Kluwer.

Skow, B. 2014. Are There Non-Causal Explanations (of Particular Events)? *British Journal for the Philosophy of Science* 65: 445–467.

Strevens, M. 2008. *Depth: An Account of Scientific Explanation.* Cambridge, MA: Harvard University Press.

Suárez, M. 2010. Scientific Representation. *Philosophy Compass* 5: 91–101.

Suppe, F. 1989. *The Semantic Conception of Theories and Scientific Realism.* Urbana, IL: University of Illinois Press.

Thomasson, A. 2006. Debates about the Ontology of Art: What Are We Doing Here? *Philosophy Compass* 1: 245–255.

Toraldo di Francia, G. 1978. What Is a Physical Object? *Scientia* 113: 57–65.

van Fraassen, B. C. 1980. *The Scientific Image.* Oxford: Oxford University Press.

van Fraassen, B. C. 1989. *Laws and Symmetry.* Oxford: Oxford University Press.

van Fraassen, B. C. 1991. *Quantum Mechanics: An Empiricist View.* Oxford: Oxford University Press.

van Fraassen, B. C. 2008. *Scientific Representation: Paradoxes of Perspective.* Oxford: Oxford University Press.

van Fraassen, B. C. 2014. One or Two Gentle Remarks about Hans Halvorson's Critique of the Semantic View. *Philosophy of Science* 81: 276–283.

van Fraassen, B. and Sigman, J. 1993. Interpretation in Science and in the Arts. *Realism and Representation*, ed. G. Levine, 73–99. Madison: University of Wisconsin Press.

Vickers, P. 2013. A Confrontation of Convergent Realism. *Philosophy of Science* 80: 189–211.

von Neumann, J. 1932. *Mathematische Grundlagen der Quantenmechanik.* Berlin: Julius Springer. Trans. by R. T. Beyer as *Mathematical Foundations of Quantum Mechanics.* Princeton, NJ: Princeton University Press, 1955.

Wallace, D. 2012. *The Emergent Multiverse: Quantum Theory according to the Everett Interpretation.* Oxford: Oxford University Press.

Wilson, J. 2017. Determinables and Determinates. *The Stanford Encyclopedia of Philosophy* (Spring 2017 Edition), ed. E. N. Zalta. https://plato.stanford.edu/archives/spr2017/entries/determinate-determinables/ (last accessed: 26.6.2023).

Wolff, J. 2012. Do Objects Depend on Structures? *British Journal for the Philosophy of Science* 63: 607–625.

Worrall, J. 1989. Structural Realism: The Best of Both Worlds? *Dialectica* 43: 99–124 (also in D. Papineau (ed.), 1996: *The Philosophy of Science.* Oxford: Oxford University Press, 139–165).

Michela Massimi

The Perspectival Nature of Scientific Representation

Abstract: In this paper, I go back to Bas van Fraassen's seminal discussion of perspectivity in the context of scientific representation as a springboard for distinguishing between two possible ways of locating the perspectival nature of the representation. I will do so by drawing an analogy between art and science, as per my recent book *Perspectival Realism* (2022). The perspectival nature of the representation challenges some deep assumptions about how scientific models represent reality; and, ultimately, I suggest, it creatively points the way to new possible directions for rethinking realism in science.

> In the notion of perspective, as so often, we have a cluster concept, with multiple criterial hallmarks. There is no defining common set of characteristics, only family resemblances among the instances. Whether or not something is aptly called perspectival depends on whether some appropriate subset of these hallmarks are present, but what amounts to 'appropriate' we cannot delimit precisely either. (Bas van Fraassen 2008, 59)

1 Preamble

It is a great honour and a real joy to be invited to contribute to this volume, which celebrates Bas van Fraassen's profound and multifaceted contributions to philosophy of science. Very few people have shaped and defined our field as Bas van Fraassen has done in the last 50 years or so. His work has been hugely influential for my generation, and for my own intellectual trajectory since I was a graduate student. The themes of *The Scientific Image* shook some deeply instilled realist intuitions I have always had and gently invited me to reconsider the issue of realism through broadly empiricist lenses. *Laws and Symmetries* inspired me to look at the topic of laws of nature as a treasure throve of philosophical ideas. But, most of all, it was his book on *Scientific Representation: Paradoxes of Perspective* that set me on a journey I am still on.

It was not just the intriguing analogy between art and science in *Scientific Representation* that captivated my attention, or the masterful scholarship of tracing back to Kant, Carnap, and Putnam some central themes. What I have always found most inspiring about that book was the way it posed some genuinely interesting philosophical questions about what it is like for science to represent reality —questions that are, for their own nature, very much open-ended and bound to

https://doi.org/10.1515/9783111019802-008

remain such. This is philosophy at its best. Philosophy as is and ought to be. Not the kind of philosophy that aims at proselytizing, or offering "knock-down arguments" against rival philosophical positions. But instead, the kind of philosophy that opens up unexpected and intellectually rewarding journeys of exploration.

2 Locating the Perspectival Nature of Representation

In his landmark book, *Scientific Representation: Paradoxes of Perspective* (2008) van Fraassen crisply laid out some important hallmarks that make a representation "perspectival":

> *occlusion, marginal distortion, texture-fading (grain), angle,* and, with special importance, *explicit non-commitment* and the 'horizon of alternatives'. These are all characteristics that relate to the content of the representation. (van Fraassen 2008, 59, emphases in the original)

Occlusion concerns the shape and size of the object being represented *from a particular vantage point* and marginal distortion is the outcome of representing the object from that vantage point. Perspectographs were machines used among Renaissance painters to produce perspectival representations. Any given object would have been seen through a pinhole in a plain frame, with the perspectival lines subtended to the angle of aperture, resulting in a marginal distortion of the image. Texture-fading is evident in the use of colours in perspectival representations to coarse-grain details as the eye moves from the foreground to the background of the picture. The angle from which the scene is being depicted determines and affects these other effects.

But the two most important hallmarks in the cluster concept of perspectival representation are *explicit non-commitment* and *horizon of alternatives.* Van Fraassen refers to Dominic Lopes's notion of *explicit non-commitment* (van Fraassen 2008, 38–39) to explain how a perspectival representation may represent its subject as having some property (or properties) that will preclude being committal in some other respect. In other words, it is a hallmark of perspectival representation that there is no God's-eye view, no overarching perspective to which it may be possible to reduce or bring under different perspectival representations:

> There is still in each case a choice to represent the subject as thus or so, and this precludes representing it as having certain other properties, which could have been selected for depiction in another picture in the same style. (van Fraassen 2008, 39)[1]

One is here reminded of earlier remarks in *The Empirical Stance* (2002, 148–151) concerning how interpretation enters into art as well as in science in that "the facts leave sufficient leeway to allow for representation as A or as B, where A and B are incompatible (in that nothing could at the same time be both A and B" (2002, 148). Consider as an example Diego Velázquez's famous painting *Las Meninas* (1656). The painting represents the Infanta Margarita, right in the centre of the picture, as curiously gazing at the spectator while being dressed. The scene is painted from the point of view of the royal couple facing and looking at the Infanta Margarita: their presence in the scene can be inferred from the mirror image at the back of the room. Velázquez might have equally represented the Infanta as a toddler playing with the dog and the parents smiling at her.

This example illustrates a first possible way of thinking about what makes a representation "perspectival" (I am going to label it *perspectival*$_1$ *representation*). Perspectivity can be thought of as affecting the *content* of the representation: the painting *represents* the Infanta Margarita and her parents *as* a composed and timid five-year-old girl watched by her parents. But the painting is explicitly non-committal as to whether, for example, the three-year-old Margarita was (or was not) a playful toddler cheered by her parents.

The situated, partially occluded, and explicitly non-committal nature of perspectival representation points to the need for a broader viewpoint, one which —barring any view from nowhere—can at the same time encompass different points of view, without reducing, merging, or overlapping them. Van Fraassen calls this broader viewpoint the "horizon of alternatives":

> When we think of a picture as being drawn from one point of view (the location of the eye and direction of vision), we are attending to its alternatives: thinking of it as set in a 'horizon' of other perspectives on the same objects. [...] Thus, as to the hallmarks of perspective: the characteristics listed [occlusion, grain, non-committal] will not suffice if applied 'piecemeal'. The content of the picture must be related to a 'horizon' of alternatives that we can think of as

1 Catherine Elgin returns to the *explicitly non-committal* nature of perspectival representation in her book *True Enough*, where she observes that "A representation is *explicitly noncommittal* with respect to a given property if its representing the having of one feature precludes its taking a stand on the having or lacking of another. A representation of a man wearing a hat is explicitly noncommittal with respect to whether he is bald, because representing him as hatted makes it impossible for the picture to commit itself on the question of his baldness. [...] because of these limitations, there cannot be a single, comprehensive perspectival representation that represents everything from a single point of view. The 'God's eye view' cannot be a point of view" (2017, 154–155).

> coming from 'different points of view', if these characteristics are to count as marks of perspectivity—and the explicit or implicit reference to such a horizon of alternatives is what is most important in the concept of perspectivity. (van Fraassen 2008, 39)

The accompanying footnote appended to this quote refers to the work of Nelson Goodman and Catherine Z. Elgin on pictorial symbol systems. And Elgin herself has returned to this topic in her book *True Enough* (2017), with a slightly different terminology. For instead of "horizon of alternatives" she refers to a "logical space" as a "multidimensional array of possibilities" where to

> locate an item in a logical space is to determine which of the possibilities defined by that space it realizes. [...] Representations in a logical space, like representations in a physical space, can be perspectival. They can show how occupants of that space appear from a certain vantage point. [...] I take the term 'perspectival' to refer to any representation that represents how things appear from a particular point of view. (Elgin 2017, 155)

Out of the artistic metaphor, perspectival representations are often deemed to be representations *from a particular point of view* (*perspectival*$_1$), where the particular point of view is often taken to affect in relevant ways the *content* of the representation itself. This is well expressed by the familiar notion of *representing-as* (see Elgin 2017, 252–257; Hughes 1996; Frigg 2010), which has often been contrasted with the notion of *representation of* in the literature. In brief, *x* is a representation *of y* insofar as *x* denotes *y: Las Meninas* is a representation *of* the Spanish royal family because it denotes it (by fiat or by stipulation). But *Las Meninas* is an *x*-representation because it represents the Spanish royal family *as* (for example) a composed and rather subdued family.

Representation-as captures, in my view, some salient features of the perspectival$_1$ nature of the representation. Not only is the scene *represented from* a particular point of view that—as such—affects the *content* via the occlusion caused by objects in the foreground; the angle subtended to the spectator's eye; the marginal distortion of the resulting representation; the coarse-graining of the sketchy and idealized image, and so on. But its *representational content* is also *represented-as y* (rather than as *z*, or as *k*, or as *j*) *from* that vantage point, where *y*, *z*, *k*, or *j* are all attributes or properties belonging to the "horizon of alternatives" that each situated representation (qua *representation-as*) respectively ascribes or imputes to the target. These attributes or properties may well be incompatible with one another, and this incompatibility is usually accepted as unproblematic in artistic representations, where an element of interpretation is inevitably present.[2]

2 Van Fraassen clarifies this point in *The Empirical Stance* (2002, 148–149): "The painting represents the subjects as arrogant or as complacent, and the fact is that their comportment, as dis-

When it comes to perspectival$_1$ representations in science, philosophers of science (and I want to clarify for avoidance of any doubt that van Fraassen is not one of them) have sometimes thought of them in terms of many models for the same target system. Consider, for example, models of the atomic nucleus (see Margaret Morrison 2011 for a discussion). The shell model of the nucleus *represents* the target system *as* a set of concentric shells when viewed from the point of view of phenomena such as e.g. radioactive decay chains. But when viewed from the point of view of phenomena concerning, say, quantum chromodynamics, the quark model best represents the nucleus as a bunch of valence quarks exchanging gluons. And when viewed from the point of view of stellar nucleosynthesis, the cluster model best represents the nucleus as a tetrahedron where nucleons are clustered in equal and even numbers. And from the point of view of yet other phenomena such as e.g. nuclear fusion, the liquid drop model best represents the nucleus as a drop of incompressible nuclear fluid. In this example, each family of models can be regarded as representing selected features of the target system *as y*, or *as j*, or *as k*, where *y*, *j*, and *k* are different properties often enough incompatible or inconsistent with one another. The perspectivity of the representation afforded by each model is cashed out in terms of property-ascription to the target system, under this reading.

This way of thinking about the perspectival nature of scientific representations afforded by scientific models raises some interesting challenges when trying to explain what good (if any) is a plurality of scientific models for the same target system. If all there is to perspectivism as a philosophical position is the view that models afford perspectival$_1$ representations, it would seem that perspectivism is at best irrelevant and unstable as a view, at worst incoherent (see Chakravartty 2017, Section 6.3). Unsurprisingly, critics of perspectivism have cast doubts on the ability of perspectivism so understood to deliver on realism, for example.

But perspectivism is not necessarily a stumbling-block to realism, I maintain. Indeed, there is a lot more to perspectivism as a philosophical view, and teasing out the intricacies of the view may prove a wonderfully rewarding intellectual journey, like the one I embarked on about ten years ago and whose main results were published in Massimi 2022. For example, to rescue perspectival representation from the problem of inconsistent models (see Massimi 2018 and 2019 for a discussion), and to further develop the very important insights from van Fraassen and Elgin

---

played to the painter, allowed both interpretations. So at the very least we must think of the representations of nature that sciences gives us as representations of nature as thus or so. The 'as' indicates an interpretive element, and our example shows that the facts leave sufficient leeway to allow for representations as A or as B, where A and B are incompatible (in that nothing could at the same time be both A and B)."

about the "horizon of alternatives" in a different direction, I have suggested that we think of the representation afforded by scientific models along less familiar —yet no less relevant—lines, namely along the lines of what in Massimi (2022), Chapter 2, I referred to as *perspectival*$_2$ representation (masterfully displayed by Jan van Eyck's 1434 *Arnolfini Portrait* in art).

Think of perspectival representation not so much in terms of how the content is represented *from a specific vantage point as* having some property *y* rather than *z* or *k*, among rival possible ones. Think of it instead as opening up a "window" on reality, to use the word of the art historian Erwin Panofsky (1991, 55), so to speak, that extends well beyond the boundaries of the representation itself and where the depth, angle, and scale of the representation leave us enough room to *make inferences* about the space and what's in it (e. g. Are there two people witnessing the Arnolfini betrothal? Who are they? Is there another room opening up behind them?). Here is an alternative possible way of understanding the perspectival nature of the scientific representation:

Perspectival$_2$ *Representation:*

(a) The representation is not just situated. It has also a directionality. It is *towards* a vanishing point that creates the perspectival illusion of a space extending beyond the boundaries of the represented scene, making visible what is technically not visible in the represented content.[3]

(b) The representational *content* is not itself perspectival in that it is not affected by the vantage point *from* which the representation takes place: it is not an instance of *representing-as* understood as ascribing alternative and incompatible attributes or properties. Van Fraassen's hallmarks still apply, but the 'horizon of alternatives' is put to a different use: it allows for a plurality of modal inferences about the target system.

I have explicated the full implications of the notion of perspectival$_2$ representation for how to rethink the connection between realism and perspectivism in Massimi (2022). In the rest of this paper I simply attend to a more general problem that seems to arise from the perspectival nature of representations in science.

3 This was the greatest achievement of Renaissance perspectival art, according to the art historian Erwin Panofsky. Perspective transformed "the entire picture […] into a 'window', […] we are meant to believe we are looking through this window into a space [where] […] all perpendiculars […] meet at the so-called central vanishing point, which is determined by the perpendicular drawn from the eye to the picture plane. […] The surface is no longer the wall or the panel bearing the forms of individual things and figures, but rather is once again that transparent plane through which we are meant to believe that we are looking into a space, even if that space is still bounded on all sides. […] The picture has become a mere 'slice' of reality, to the extent and in the sense that *imagined* space now reaches out in all directions beyond *represented* space" (Panofsky 1991, 55 and 60–61).

3 A Problem with Perspectival Representations in Science

Intriguing as it might be, the analogy between art and science has its limits. In what sense, if any, one might ask, can a scientific model of, say, the nucleus, a protein, or something else be labelled as perspectival as my earlier analogies with Velázquez's *Las Meninas* or van Eyck's *Arnolfini Portrait* implied? Van Fraassen himself (2008, 86) warns us about possible disanalogies between art and science: "general scientific theories, in their 'official' formulation, are not perspectival descriptions and their models—if we consider the entire range of models for a given theory—are generally not perspectival representations". Van Fraassen had already clearly articulated this point in *The Empirical Stance* (2002, 148–151). There, referring to Newtonian's celestial mechanics, he envisages a critic—a "conservative philosopher"—who might reply that "science, like art, interprets the phenomena, and not in a uniquely compelled way. Science itself, however, does not admit of alternative rival interpretations" (van Fraassen 2002, 149).

Although the phenomena did not compel Newton's interpretation (to the extent that a rival one by Einstein was eventually adopted), science (by contrast with art) does not easily lend itself to be interpreted in a plurality of rival ways. Against the "conservative" philosopher, van Fraassen rejoins that the "history of science put the lie to this story, and in successively more radical ways", for "in science too, we find interpretation at two different levels. The theory represents the phenomena as thus or so, and that representation itself is subject to more than one tenable but significantly different interpretation" (van Fraassen 2002, 151).

From models of measurement to data models; from phenomenological models to theoretical models; from diagrams, histograms, charts, all the way up to more abstract models: our images of nature are inevitably always situated, from a specific vantage point that affects the content of the representation itself (as when we measure the temperature in Kelvin rather than in Celsius; or the length in metres rather than feet; or the passage of time with atomic clocks rather than sundials). Scientists know how to reliably use instruments, measurement techniques, and models to deliver perspectival representations, just like Renaissance artists knew how to reliably use a perspectograph to draw in perspective.

The problem with perspectival representations concerns how to explain *our experience* of a three-dimensional perspectival space that extends well beyond the boundaries of the canvas and its flat two-dimensional drawings. How can the experience of a three-dimensional space that extends indefinitely be so skilfully, so ingeniously, yet so illusorily produced by the vanishing point of the mirror in the *Arnolfini Portrait?* How is it that from perspectival representations in science (be they data models—histograms, charts, diagrams—models of measurements, phenomenological and theoretical models), we *can experience* a world of electrons, Higgs bosons, neutrinos, proteins, eukaryote cells, DNA strands, and so on and so forth? In other words, why is it that *in spite of* the perspectival$_1$ nature of the scientific representation—which is always situated, always from a vantage point that affects the content of the representation via occlusion, marginal distortion, and so on—we can nonetheless legitimately and arguably claim to experience a "window on reality"?

To a critic, who might hasten to reply on the ground that what we experience is in fact just an illusion—as illusory as the space masterfully created by van Eyck's mirror—should not we reply that perspectival as they might be, these scientific representations are in fact representations of a world (or slice thereof) that exists independently of us? A world that is as real as is the tree visible from the window in the *Arnolfini Portrait,* no matter if we do not have a way of coming to know it outside the boundaries of van Eyck's representation itself? Is not the natural world teeming indeed with electrons, Higgs bosons, neutrinos, proteins, DNA strands, eukaryote cells, hellebores, and chemical elements, regardless of how perspectival our scientific representations of them might all be?

Philosophers of science with realist leanings have long argued that the perspectival nature of the scientific representation is compatible with realism in science.[4] Philip Kitcher (2001, 51), for example, has warned against conflating the possibility of constructing representations with that of constructing the world. Our selected vantage point—if understood, for example, as the selective language with its taxonomic categories that we use to describe "the world *that we care about*" (Kitcher 2001, 46 emphasis in original)—is just one among many possible ones:

> Different ways of dividing nature into objects will yield different representations of reality. Users of different schemes of representation may find it difficult to coordinate their languages. Properly understood, however, the truths they enunciate are completely consistent. (Kitcher 2001, 47)

4 For a non-exhaustive and rather diverse list of examples in this direction, see Psillos (2012) for a semantic analysis in terms of causal descriptivism; Chakravartty (2011) for a defence of ontological pluralism; Teller (2019) for the articulation of a form of perspectival realism that takes the problem of reference-continuity at its heart; Cartwright (1999), Dupré (1993), and Chang (2012) for a variety of realist commitments that focus on a metaphysics of causal powers, or on taxonomic pluralism; French (2014) and Ladyman and Ross (2007) for sustained articulations of structuralist responses to a similar problem.

Navigating between the realist stance articulated by Kitcher and the irreducibly perspectival nature of scientific representation established in van Fraassen's wake leads me to *perspectival realism*. My monograph (Massimi 2022) is my own personal journey through this fascinating and pretty much still open-ended philosophical landscape without any claim, intention, or expectation of ever exhausting the myriad of aspects and nuances it carries with it.

My journey toward that position began with a problem concerning the perspectival nature of the scientific representation and realism in science. The problem is summarized by two distinct assumptions pulling in opposite directions. Let us call them the *representationalist assumption* and the *perspectivalist assumption:*

> *The representationalist assumption.* Scientific models (partially) represent relevant aspects of a given target system *S*. A scientific model is true (or approximately true) when the model provides a partial yet accurate representation of the target system.

A couple of qualifications are in order. First, it might seem like a slip of tongue to say that a scientific model is true—after all, scientific models under the semantic view are not propositional objects amenable to being true or false. While this is of course correct, the question about representation tends to go hand in hand with the question about truth in science.

We use scientific models—and the representations they afford—to advance *claims* about relevant aspects of the world. Some of these model-based claims are deemed true so long as the model is deemed to provide a partial yet accurate representation of the target system. Thus, it might be tempting to understand the question about truth in terms of the accuracy of the representation afforded by the scientific model.

The second qualification is related. The representation afforded by any scientific model can only be approximately true. Being approximately true means that each model can accurately represent only some parts or portions of the target system while abstracting and idealizing others. For example, the liquid drop model of the nucleus may provide an approximately true story of how the binding energy can be released in nuclear fission, while distorting the representation of the atomic nucleus in other respects.

Realism in science has typically been associated with the expectations that scientific models (and scientific theories as families of models) "accurately-albeit-partially" represent relevant aspects of the target system within the limits of idealization and abstraction. Some of the problems in cashing out perspectival realism originate, in my view, from rather stringent readings of the representationalist assumption. Relaxing those stringent readings is not tantamount to foregoing realism in science. At most, it is foregoing a specific brand of realism (dispositional essentialist in flavour). But for now, it suffices to say that the representationalist assumption is in tension with another assumption that is often at play in discussions of scientific modelling, namely:

> *The perspectivalist assumption.* Scientific models offer perspectival representations of relevant aspects of a given target system *S*. This is (often implicitly) understood along the lines of *perspectival$_1$ representation* rather than *perspectival$_2$ representation*. Namely, scientific model M_1 *perspectivally$_1$ represents* S *as* z; but model M_2 *perspectivally$_1$ represents* S *as* y (where *z* and *y* are incompatible or even inconsistent properties belonging to a "horizon of alternatives").

The perspectivalist assumption—if understood primarily along the lines of perspectival$_1$ representation—suggests that rival and often incompatible property ascriptions are at play for the same target system S across several models. Key to this assumption is the idea that models offer representations that are perspectival because they introduce into the target system S rival or even incompatible properties out of a range. *This* model *here* represents the nucleus *as* a bunch of quarks exchanging gluons; but *this other model* represents the nucleus *as* a set of concentric shells which nucleons fill in magic numbers (2, 8, 20, 28, 50, 82). Examples of this nature are familiar and ubiquitous in science.

The tension between these two assumptions underlies the problem of perspectival representation, in my view. Being able to assert that the world *is* thus or so would seem to be precluded within the bounds of our perspectival$_1$ representations. This chimes with varieties of anti-realist views and arguments that I will not rehearse or review in what follows because my task ahead is a constructive one, to show how the problem of perspectival representation can be addressed, if not resolved altogether, as soon as some stringent readings of both the representationalist and the perspectivalist assumption are scrutinized and ruled out—see Massimi (2018) for some initial steps in this direction and for a book-length analysis see Massimi (2022).

But, in the rest of this chapter, I'd like to say something about a possible way of deflating the risk associated with perspectival$_1$ representation and property-ascription inherent in *representing-as*. Perspectival representations do not just display the aforementioned hallmark criteria (i. e. occlusion, marginal distortion, angle, grain, explicit non-commitment, and horizon of alternatives). They are also indexical (to use again one of van Fraassen's terms): they are *used* by agents for specific intents, goals, and purposes.

Scientific representations represent the target system *from a specific vantage point* as that vantage point relates to an agent who is *using* the representation *for a specific goal*. Does not this appeal to the indexicality of the representation deflate (or at least mitigate) the problem of perspectival representation? Are not perspectivity and indexicality pragmatic (as opposed to epistemic) features of scientific representation?

4 Indexicality: The Pragmatics of Perspectival Representations

Indexicality refers to the context-sensitive *use* that epistemic agents like us make of perspectival representations. As van Fraassen rightly observes:

> the activity of representation is successful [...] only if the recipients are able to receive that information through their "viewing" of the representation. [...] The recipient must be in some pertinent sense able to relate him or herself, his or her situation, to the representation. (van Fraassen 2008, 76 and 80)

Measurements, maps, models, all offer perspectival representations for specific *uses*, goals, purposes of agents. Van Fraassen (2008, 176) distinguishes between "*Measuring is Locating* and *Measuring is Perspectival*", whereby "what is perspectival is not the action of measuring but the content of the measurement outcome, and locating is an action, not a content". The two notions are intrinsically connected: "the measurement outcome does after all represent the target as located in a certain logical space" (van Fraassen 2008, 176). The action of representing is indexical because it locates the perspectival content of the representation in a logical space for an agent to use it for specific purposes. Representing requires that the agent locates themselves with respect to the horizon of alternatives so as to see the representation *from a specific vantage point* and to *use* it for specific purposes. Interesting consequences derive from this distinction.

Consider, for example, measurements as a kind of perspectival$_1$ representation. Room temperature can be measured on either the Fahrenheit scale or the Celsius scale. Yet it would be hasty to conclude that these two measurement scales locate the same object in two different logical spaces. For there is a "single logical space in which the two scales are, in effect, two coordinate systems" (van Fraassen 2008, 174) and a linear transformation allows the agent to seamlessly move from one to the other. Although we might say that the Fahrenheit thermometers and the Celsius thermometers offer "different perspectives on the same magnitudes", this would only be a *façon de parler*, rather than a genuine difference, since the underlying theory is the same.[5]

5 Cf. van Fraassen (2008, 173): "If a thermometer is used by one person to locate the air in his room

Along similar lines, Ron Giere (2006, Ch. 3) discusses what I call the perspectival$_1$ representations associated with measuring the intensity and distribution of gamma rays coming from the centre of the Milky Way "*as indicated by COMPTEL or OSSE*" (Giere 2006, 48 emphasis in original) or other instruments and their measurement outcomes, where again the perspectival content afforded by each of these different measurement techniques and instruments is functional to specific uses and purposes of the agents.[6] This view is in accord with Giere's overarching agent-based conception of scientific representation (Giere 2010), whereby representation is understood as a four-step process: agents (1) intend (2) to use model M (3) to represent a part of the world W (4) for some purpose P. How should one think of perspectivity and indexicality in relation to scientific models? Like van Fraassen and Giere, Elgin too, stresses the pragmatics of perspectival representations:

> Science is not something that just happens to us; it is something we do. To do it, we need to use scientific representations. Use is a pragmatic matter; and to make use of anything an agent needs to properly locate herself with respect to it. This is evident in the relation between science and technology. To develop an alloy that resists metal fatigue, a metallurgist has to be able to recognize signs of metal fatigue. To do that, she must adopt a perspective from which she can discern metal fatigue is present. (Elgin 2017, 157)

Elgin, too, goes on to note that the

on the Fahrenheit scale, and by another to locate it in the space of possible mean kinetic energies of its molecules, the two are locating the same thing, by means of the same instrument, in two different logical spaces. That we should not at once call a change of perspective, though it certainly marks a change in ways of thinking about the air. What if a theory then equates temperature and mean kinetic energy? In that case we should say that *relative to the theory* it is appropriate to call this a change of perspective. This qualification is important, though it may be left tacit in a context where the theory has been entirely accepted."

6 Giere (2006, 48, emphasis in original): "Humans and various other electromagnetic detectors respond differently to different electromagnetic spectra. Moreover, humans and various other electromagnetic detectors face the same spectrum of electromagnetic radiation and yet have different responses to it. In all cases, the response of any particular detector, including a human, is a function of *both* the character of the particular electromagnetic spectrum encountered and the character of the detector. Each detector views the electromagnetic world from its own *perspective*. Every observation is perspectival in this sense." And in the associated footnote Giere specifies: "Instruments interact directly with objects in the world. They typically *produce* representations of the objects in question. But the representation is a *product* of the interaction between instrument and object, not part of the process by which the instrument detects aspects of the objects" (Giere 2006, fn. 7, 126 emphasis original).

> only way for an observer to perform the experiments or make the measurements is to locate herself in the logical space of the phenomena. Thus her stance is indexical and perspectival. [...] Measurement is always indexical and perspectival. Hence for science to do its epistemic job, it must involve indexical, perspectival representations. [...] Those representations are objective in that they contain information that is invariant across representations of the same object. They are testable in that multiple representations of the same objects from the same perspective yield equivalent information, and in that information can be accessed from multiple perspectives. (Elgin 2017, 158–159)

As a case in point, Elgin mentions multiple models of the atomic nucleus, whereby

> in some significant respects the nucleus behaves like a liquid drop, and another model highlights that in some other significant respect it behaves as though it has a shell structure. [...] There is no reason why the same thing should not share some significant properties with liquid drops and other significant properties with rigid shells. It may be surprising that the same thing could have both sets of features, but there is no logical or conceptual difficulty. The models afford different perspectives on the same reality. (Elgin 2017, 270)

A common theme runs through indexicality as a pragmatic feature of perspectival$_1$ representation: namely that the ability of the agents to locate themselves in the logical space of alternatives and to pragmatically use a multitude of (presumably) perspectival$_1$ representations for different purposes is not only necessary for testing and measuring in science, but it is also metaphysically and epistemically innocent.

It is metaphysically innocent because it does not affect reality itself. Perspectival$_1$ representations—if understood in pragmatic terms—do not engender a multitude of perspectival facts, somehow reminiscent of Kuhnian *Gestalt switches* and "living in a new world".

Moreover, indexicality seems to be epistemically innocent. Perspectival$_1$ representations presumably share the same underlying theory;[7] or yield information that can be accessed from multiple perspectives;[8] or are subject to linear transformations that allow agents to seamlessly move from one perspective to another.

---

7 Perspectival$_1$ representations might be thought of as representational models within a single underlying theory, e.g. the pendulum model and Galileo's inclined plane model for free fall within Newtonian mechanics (see Giere 2006, 61–62).

8 I take this to be Elgin's own view about why shell models and liquid drop models for the nucleus do not constitute a logical or conceptual difficulty. As I read Elgin, her view takes models as depending on what she calls "exemplification": "when *x* represents *y* as *z*, *x* is a *z*-representation that *as such* denotes *y* [...] in being a *z*-representation, *x* exemplifies certain properties and imputes those properties or related ones to *y*" (Elgin 2017, 260). Thus, perspectival$_1$ representations (e.g. the shell model and the liquid drop model) might exemplify different sets of properties each of which might be displayed by the represented target system in the relevant context for specific purposes: "Sometimes, although the target does not quite instantiate the features exemplified by the model, it

I think the point about metaphysical innocence stands. But I am less sanguine about the point on epistemic innocence in situations where:

- the partiality of the representations at issue concerns incompatible or even inconsistent property ascriptions to the target system;
- the representation is not just intended to *exemplify* or be *similar* in some relevant respects to selected aspects of the target system (while being false in others); but instead, it is intended to ascribe to that target system property *y* rather than property *z* (where those properties are incompatible or inconsistent with one another);
- there is no seamless transformation across perspectives or single unifying theory behind the plurality of perspectival$_1$ representations.

Indexicality seems to me to leave the epistemic problem untouched.[9] However, indexicality conceals two very important insights. First, the *use* that particular communities make of their perspectival$_1$ representations for particular purposes is absolutely key to understanding the *local* and always *situated* nature of scientific knowledge claims. Second, the very action of representing perspectivally is directed to a specific goal (or vanishing point, to return to the metaphor from art). The answer to the problem of how wonderfully diverse human beings like us come to reliably know the world from a plurality of situated scientific perspectives ultimately lies in how one understands the nature of perspectival$_2$ representations delivered by scientific models. These are the issues that I have endeavoured to discuss by articulating my view on perspectival realism (see Massimi 2022).

As announced at the beginning, van Fraassen's work on *Scientific Representation: Paradoxes of Perspective* set me on a fascinating intellectual journey. The final outcome of this journey across a variation of a Kantian theme (Massimi 2017) is, as I see it, to live up to the promise of realism in science. It is, however, a very different kind of realism from more traditional ones that back in 1980 Bas van Fraassen, rightly in my view, warned us against. The final outcome is instead a kind of real-

---

is not off by much. Where their divergence is negligible, the models, although not strictly true of the phenomena they denote, are true enough of them. […] Where a model is true enough, we do not go wrong if we think of the phenomena as displaying the features that the model exemplifies. Whether a representation is true enough is a contextual question. A representation that is true enough for some purposes or in some respects is not true enough for or in others. This is no surprise" (Elgin 2017, 261).

9 Intentionally so, of course, in van Fraassen's empiricist anti-realist stance and similarly in Elgin's defence of models as "felicitous falsehoods". The problem of perspectival representation is a problem for the realists at heart who nonetheless take seriously the situated and perspectival nature of our scientific knowledge.

ism *within the bounds of our perspectival representations*—the only kind of realism that I think we can legitimately reclaim as our own. I owe to the towering work of Bas van Fraassen to have opened up for me a wonderful world of philosophical possibilities to explore over the past two decades.

Acknowledgement: This paper is an abridged version of Chapter 2 "The Perspectival Nature of Scientific Representation" from my monograph *Perspectival Realism* (OUP, 2022). All rights reserved. Reprinted by permission of the publisher, © 2022 Oxford University Press. I am very grateful to Bas van Fraassen for his work and conversations over the years that have been inspiring and instrumental in making me think about this topic and questions. Many thanks to the audiences at the Lauener Symposium in Bern (Sept. 2018), Boston Colloquium in Philosophy of Science (April 2019) and Workshop on *Idealisations across the Sciences* (Prague, June 2019) where earlier versions of this paper were presented. This paper is part of a larger project that received funding from the European Research Council (ERC) under the European Union's Horizon 2020 research and innovation programme (grant agreement European Consolidator Grant H2020-ERC-2014-CoG 647272 *Perspectival Realism. Science, Knowledge, and Truth from a Human Vantage Point*).

References

Cartwright, N. 1999. *The Dappled World.* Cambridge: Cambridge University Press.

Chakravartty, A. 2011. Scientific Realism and Ontological Relativity. *The Monist* 94(2): 157–180.

Chakravartty, A. 2017. *Scientific Ontology.* New York: Oxford University Press.

Chang, H. 2012. *Is Water H_2O? Evidence, Realism and Pluralism.* Dordrecht: Springer.

Dupré, J. 1993. *The Disorder of Things: Metaphysical Foundations of the Disunity of Science.* Cambridge, MA: Harvard University Press.

Elgin, C. Z. 2017. *True Enough.* Cambridge, MA: The MIT Press.

French, S. 2014. *The Structure of the World.* Oxford: Oxford University Press.

Frigg, R. 2010. Fiction and Scientific Representation. In: *Beyond Mimesis and Convention*, eds. R. Frigg and M. C. Hunter, 97–138. Dordrecht: Springer.

Giere, R. 2006. *Scientific Perspectivism.* Chicago: University of Chicago Press.

Giere, R. 2010. An Agent-Based Conception of Models and Scientific Representation. *Synthese* 172: 269–281.

Hughes, R. I. G. 1996. Models and Representation. *PSA: Proceedings of the Biennial Meeting of the Philosophy of Science Association* 1996, Vol. 2: 325–336.

Kitcher, P. 2001. *Science, Truth, and Democracy.* Oxford: Oxford University Press.

Ladyman, J. and Ross, D. 2007. *Everything Must Go: Metaphysics Naturalized.* Oxford: Oxford University Press.

Massimi, M. 2017. What Is This Thing Called 'Scientific Knowledge'? Kant on Imaginary Standpoints and the Regulative Role of Reason. *Kant Yearbook* 9/2017: 63–83.

Massimi, M. 2018. Perspectival Modelling. *Philosophy of Science* 85: 335–359.

Massimi, M. 2022. *Perspectival Realism.* New York: Oxford University Press.
Morrison, M. 2011. One Phenomenon, Many Models: Inconsistency and Complementarity. *Studies in History and Philosophy of Science* 42: 342–351.
Panofsky, E. 1991. *Perspective as Symbolic Form.* Cambridge, MA: MIT Press.
Psillos, S. 2012. Causal Descriptivism and the Reference of Theoretical Terms. In: *Perception, Realism, and the Problem of Reference*, eds. A. Raftopoulos and P. Machamer, 212–238, Cambridge: Cambridge University Press.
Putnam, H. 1990. *Realism with a Human Face.* Cambridge, MA: Harvard University Press.
van Fraassen, B. 2002. *The Empirical Stance.* New Haven: Yale University Press.
van Fraassen, B. 2008. *Scientific Representation: Paradoxes of Perspective.* Oxford: Oxford University Press.

Part II: Constructive Empiricism and Other Foes of Scientific Realism

Finnur Dellsén

Expanding the Empirical Realm: Constructive Empiricism and Augmented Observation

Abstract: Manifestationalism holds that science aims only to give us theories that are correct about what has been observed thus far. Several philosophers, including Bas van Fraassen, have argued that manifestationalism cannot make sense of the scientific impetus to make new observations, since such observations only risk turning manifestationally adequate theories into inadequate ones. This paper argues that a strikingly similar objection applies to van Fraassen's own constructive empiricism, the view that science aims only to find theories that are empirically adequate. Roughly, the objection is that constructive empiricism cannot make sense of the scientific impetus to expand the limits of what can be observed, since such expansions only risk turning empirically adequate theories into inadequate ones.

1 Introduction

Any empiricism worthy of its name makes a distinction between the empirical and the non-empirical, prioritizing the former over the latter. Different forms of empiricism draw the line in different ways. Logical empiricists made a linguistic distinction between observational and theoretical *terms*, for instance, while Bas van Fraassen (1980) makes a distinction between observable and unobservable parts of the *world*, e.g. entities or processes. Whether a part of the world counts as observable depends, for van Fraassen, on whether the relevant scientific community would be able to observe it directly in the right circumstances. This relativizes observability to a given community: Something may count as observable relative to a community with greater observational powers while counting as unobservable relative to a community that is more observationally limited. To say that something is observable is, for van Fraassen, simply to say that it is observable relative to *our* community.

One consequence of this relativity in what counts as observable is that the line between observable and unobservable would shift if there are sufficiently substantial changes to the observational powers of the relevant community. An entity that was previously unobservable could become observable, not because of any real changes in the entity itself but because the relevant community has in-

https://doi.org/10.1515/9783111019802-009

creased its abilities to observe entities of this kind.[1] In this paper, I argue that this malleability in what counts as observable according to van Fraassen creates a problem for van Fraassen's constructive empiricism, which holds that scientific theories are successful in so far as they are correct about the observable aspects of the world. Roughly, the problem is that constructive empiricism seems to imply that scientists who augment their observational powers are risking that a previously successful theory might thereby become unsuccessful. Since scientists themselves do not seem to recognize this as a genuine risk—scientists do not hesitate to augment their observational powers whenever possible—there seems to be a tension between constructive empiricism and established scientific practice.

In what follows, I formulate what I consider to be the most plausible version of this argument, and consider some ways in which a constructive empiricist might respond. At the end of the day, I am not sure whether this argument constitutes a genuine problem for constructive empiricism, but as far as I can tell there is no obvious or easy response available on constructive empiricism's behalf. Constructive empiricism has been heavily criticized on numerous occasions, but I believe that many—perhaps most—of the objections that have so far been made against constructive empiricism are based on misconceptions of what constructive empiricism is and entails.[2] I am confident that the argument I consider here, whatever its merits in other respects, is not of this kind.

2 Constructive Empiricism and Observability

Van Fraassen's *The Scientific Image* (1980)[3] did not just argue for a new kind of scientific anti-realism, constructive empiricism, but also proposed an influential way of understanding what the scientific realism debate is really about. According

1 Of course, the entity would "change" in the sense that it would go from having the property of being observable to a community to lacking that same property. But this is paradigmatic example of a mere "Cambridge change" (cf. Geach 1969); hence the qualified expression "real change" in the main text.

2 For example, constructive empiricism is often criticized for arbitrarily limiting what we can know to things that happen to be observable (Alspector-Kelly 2001); but in fact constructive empiricism entails nothing about what can and cannot be known (Monton and van Fraassen 2003). Another example: Some authors criticize van Fraassen for rejecting inferences to the best explanation to the truth of theories while accepting inferences to the best explanation to the conclusion that theories are empirically adequate (e.g. Psillos 1996); but in fact van Fraassen rejects both kinds of inferences—because he rejects the entire rule or schema called "Inference to the Best Explanation" (Ladyman et al. 1997).

3 Although see also van Fraassen 1985, 1989, 2001, 2007.

to van Fraassen, there are two points of contention between scientific realists and anti-realists. The first concerns the *aim* of science: Scientific realists hold that the aim of science is to give us true theories, while anti-realists hold that the aim of science is something other than truth. The second point of contention concerns what is to *accept* a scientific theory: Scientific realists hold that accepting a theory involves believing that the theory is true; anti-realists hold that one can accept a theory without believing that it is true. In this paper, I will set aside the issue of the nature of scientific acceptance (in part because I have discussed it elsewhere; see Dellsén 2017), and instead focus on the former, i. e. the debate about the aim of science.

Van Fraassen's use of the term 'the aim of science' should not be taken too literally. Science is not the sort of being that could literally be aiming for anything in the sense that, for instance, an archer aims their bow at a target. Nor is van Fraassen referring to the personal aims or goals of individual scientists, or of the majority of scientists at a given time. Rather, van Fraassen uses the term 'aim' simply to refer to the *basic criterion of success* of the scientific enterprise —an enterprise whose participants might have various different personal goals and motives. The aim of science—or of a particular science—is thus to be understood as the basic criterion of success in (the) science. So, realism and anti-realism are in part different views about the basic criterion of success in science: Realism holds that science is genuinely successful only if (or in so far as) its theories are true, while anti-realism holds that science need not give us any true theories in order to be genuinely successful.

By definition, anti-realism is the negation of realism. So, with regard to the aim of science, anti-realism is merely the negative thesis that the aim of science *isn't truth*. Sooner or later, however, any anti-realist will have to provide a positive account of what the aim of science *is*. Constructive empiricism is van Fraassen's attempt to provide such an answer in the tradition of empiricists like Locke, Hume, Carnap and Hempel. According to constructive empiricism, "[s]cience aims to give us theories which are empirically adequate" (van Fraassen 1980, 12). In other words, the basic criterion of success in science is empirical adequacy, rather than truth. But what is "empirical adequacy"? We don't need to bother here with the precise definition; suffice it to say that a theory is empirically adequate roughly when what the theory says about observable parts of the world is true.

Now, since empirical adequacy is thus defined in terms of what is *observable*, a lot hangs on how we understand this term within van Fraassen's account. As a "rough guide to the avoidance of fallacies", van Fraassen suggests that "X is observable if there are circumstances which are such that, if X is present to us under those circumstances, then we observe it" (1980, 16). In fact, this "rough guide" is not just rough but also incomplete in van Fraassen's view, because it leaves the term 'observe' undefined. According to van Fraassen, however, this is as it should be since precisely what is involved in observation is a question to be answered by empirical science rather than philosophical analysis or stipulation (Muller and van Fraassen 2008; see also Musgrave 1985; Dicken and Lipton 2006; Dicken 2009).

Let me highlight three features of van Fraassen's understanding of observability that are relevant to the present discussion. First, as I mentioned previously, observability is relative to one's epistemic community. Something that isn't observable to one community might be observable to another, depending on the observational powers of those in each community. So—to take an extreme and perhaps overworked example—if there are Martians who can observe certain entities that we cannot observe, then those entities will be observable for the Martians but not for us (cf. Churchland 1985).

Second, van Fraassen notes that "'observable' is a *vague predicate*" (1980, 16). So while there will be things that are clearly observable (e.g., tables and chairs), and things that are clearly unobservable (e.g., the exact curvature of spacetime at a given location), there will also be things in between that aren't clearly either observable or unobservable. So unless one adopts a Williamson-style epistemicism about vagueness (Williamson 1992), it will be indeterminate whether certain things are observable or unobservable.

Finally, I wish to highlight that constructive empiricism doesn't need to draw the line between what's observable and what's not in any particular place (cf. van Fraassen 2001, 163; 2008, 110). All that's really required is that there be such a line to be drawn somewhere such that the line can plausible be considered to demarcate the 'empirical' from the 'non-empirical'. Thus, for example, a constructive empiricist could (without contradicting herself) hold that entities that can only be detected using microscopes are observable provided the microscopes are widely available and reliable. In this sense, the distinction between observable and unobservable can be thought of as a 'free parameter' within constructive empiricism.

3 The Manifestationalist Challenge

With these points in place, let us turn now to a potential problem for constructive empiricism that I will call *the manifestationalist challenge*. This challenge arises as one starts to think about the motivations one might have for adopting constructive empiricism rather than scientific realism. Recall that while scientific realism says that science isn't successful until it has given us *true* theories, constructive empiricism says that success in science requires only *empirical adequacy*, i.e. roughly truth with regard to observable aspects of the world. Since empirical adequacy is weaker than truth (i.e., all true theories are empirically adequate, but not *vice versa*), constructive empiricism apparently attributes a more modest epistemic aim to science than does scientific realism. According to van Fraassen himself (see Monton and van Fraassen 2003, quoted below) this is one of the main motivations for constructive empiricism as contrasted with scientific realism: We should not attribute to science a more ambitious epistemic aim if a more modest aim will do just as well.

But as Peter Railton (1990) points out, this apparent benefit of constructive empiricism quickly turns into a problem. For if epistemic modesty is what is at issue, then it seems that an even more modest anti-realist alternative to scientific realism should be preferable to constructive empiricism, viz. what Railton (1990, 234–235) calls *manifestationalism.* Manifestationalism holds that science aims to give us theories that are correct in all of what they say about phenomena that have *actually been observed.* (We might describe such theories as 'manifestationally adequate'.) Manifestationalism could be understood as a kind of actualistic empiricism, since it essentially holds that the aim of science is restricted to having correct theories about what is observed in the actual world. By contrast, constructive empiricism holds that science also aims to give us correct theories about phenomena that *haven't been* and *won't be* observed—but *could be.*

The option of adopting manifestationalism about the aim of science seems to suggest that constructive empiricism is unstable with regards to the motivation of epistemic modesty. The argument would go as follows: In so far as we are choosing between constructive empiricism and scientific realism with reference to epistemic modesty, it seems that we should go further and accept manifestationalism rather than constructive empiricism. After all, manifestationalism is even more modest than constructive empiricism. If, however, we reject the idea of using epistemic modesty as a criterion in the debate, then it seems that we might as well be scientific realists. Either way, the fact that constructive empiricism attributes a more modest aim to science would give us no reason to be constructive empiricists rather than scientific realists. In nutshell, the challenge here is to motivate constructive empiricism without 'proving too much' and end up motivating a more extreme form of anti-realism, viz. Railton's manifestationalism.

In an influential article, Gideon Rosen (1994) has suggested a way to meet this challenge. The idea is to show that manifestationalism makes the aim of science so modest that it can no longer explain why scientists ought to make observations that could otherwise be avoided. Rosen illustrates this with an example:

> Consider, for example, an archeologist whose theory covers all the evidence so far collected about Etruscan urns. [The archeologist] has sole license to dig in the last uninspected patch of

> ground, but it is also in [the archeologist's] power to destroy the site so that no one will ever have a chance to inspect it. What should [the archeologist] do? (Rosen 1994, 162)

Rosen's question is a little unclear as it stands, because the archaeologist might have all sorts of different kinds of reasons to dig or destroy the site. But presumably the issue is what the archaeologist should do in so far as she is trying to satisfy the aim of science, i. e. its criterion of success. What should the archaeologist do *qua scientist?* Understood in this way, the answer to this question will depend on what one takes the aim of science to be.[4]

For a manifestationalist, the answer is that the archaeologist should destroy the site, because doing so guarantees that her theory will be manifestationally adequate, i. e. correct about all actually observed urns. After all, the archaeologist's theory is already correct concerning all urns that have been observed so far, and if there are no other urns to observe then her theory is guaranteed to be correct about all urns that have been and will be observed. Hence by destroying the site the archaeologist guarantees that her theory will be successful according to manifestationalism. By contrast, for a constructive empiricist, destroying the site does not in any way help make the theory satisfy science's criterion of success, since the theory will have to answer to all observ*able* entities whether or not they have actually been, or will ever be, observed. Rather, the archaeologist should dig, because that way she can test her theory, possibly replace it, and thus increase the likelihood that the theory is empirically adequate.

The point of Rosen's example is of course to illustrate that manifestationalism implies—presumably incorrectly—that science's criterion of success would be satisfied by destroying a potentially vital piece of evidence. By contrast, constructive empiricism can make perfectly good sense of why the scientist should dig rather than destroy the site. Rosen thus concludes that constructive empiricism can be thought of as "the most cautious position—the position most compatible with empiricism in the broad sense—which at the same time renders the phenomena of scientific activity fully intelligible and ratifiable" (1994, 162). The idea, then, is that we should take the aim of science to be as modest as is possible, but not so modest as to render it unable to account for what's wrong with destroying the

4 Asay and Bordner (2015, 156–158) argue that manifestationalism has no bearing on what the archaeologist should do, mainly because (they claim) manifestationalism does not "entail anything about which sorts of scientific endeavors one should engage in" (Asay and Bordner 2015, 157). Maybe so, but it does entail something about which scientific endeavours promote the aim of science, i. e. make it more and less likely that our theories are successful according to science's basic criterion of success.

site in Rosen's example. This position is echoed by van Fraassen himself in a paper co-authored with Bradley Monton:

> Consider a range of possibilities, with 'science aims to give us true theories' on the far right side, and 'science aims to give us theories which are true in what they say about what is being observed right now' on the far left side. Realists submit that attention to the practice of good science, where bold conjectures and audacious theorizing have been rewarded with much predictive success, moves us toward the right. Empiricists, who would wish for epistemic modesty in their paradigms of rational inquiry, would tend toward the left. Constructive empiricism finds an equilibrium point between the two extremes, thus respecting both desiderata. (Monton and van Fraassen 2003, 407–408)

So, to summarize, we have seen that constructive empiricism has been criticized as unstable in so far as it is motivated by a desire for epistemic modesty in our accounts of the aim of science. According to the objection, epistemic modesty should push us all the way to a more radically anti-realist position, manifestationalism. In response, Rosen argues (on van Fraassen's behalf) that constructive empiricism is not unstable because if science aimed for anything less than empirical adequacy—as manifestationalism maintains—we couldn't explain why it would be scientifically improper to destroy potentially vital pieces of evidence, such as the archaeologist's last digging site. In the remainder of this paper, I will suggest an argument analogous to Rosen's argument against manifestationalism applies to constructive empiricism itself. The argument concerns what we may want to call 'extended observability', where the limits of what could be observed by an epistemic community are expanded or increased so that previously unobservable entities become observable.

4 Extended Observability

Let me start by distinguishing two ways in which observability can be extended according to constructive empiricism's conception of observability:

> *Epistemic immigration:* The epistemic community expands to include beings that are observationally superior in some respect.
>
> *Epistemic upgrading:* Individual agents within an epistemic community increase their observational capacities.

In both cases, the limits of observability have been *extended* in the sense that the resulting a community can observe entities that were previously unobservable. Nevertheless, it is important not to confuse these two ways in which observability can be extended within a community. Epistemic immigration does not require any individual agents to increase their observational capacities; rather, the change in observability is due only to a change in who counts as members of the relevant community. By contrast, epistemic upgrading does not require that there be any changes in who counts as members of the community; rather, the change in observability is due only to changes in the members' observational capacities.

Although these two ways of extending observability have not been clearly distinguished to my knowledge, examples of both have been extensively discussed in the literature on scientific realism, anti-realism and empiricism. For instance, Paul Churchland's thought experiment of microscope-fused humanoids is a good example of epistemic immigration:[5]

> [...] consider a race of hominoid creatures each of whom is born with an electron microscope permanently in place over his left 'eye'. The scope is biologically constituted, let us suppose, and it projects its image onto a human-style retina, with the rest of their neurophysiology paralleling our own. (Churchland 1985, 43–44)

For Churchland, this thought experiment puts pressure on the distinction between observable and unobservable, because we can presumably imagine beings of this sort for any range of things that we cannot observe—making anything in principle susceptible to being observed by some beings that could enter into our community.

In response to Churchland, van Fraassen rightly points out that Churchland's argument is only a problem for those who seek a definition of 'observable' that isn't relativized to an epistemic community. Once we relativize observability to a community, we can simply say that the limits of observability would change as the community changes:

> [...] suppose we have accepted these humanoids as persons, as members of our epistemic community. In that case we have already broadened the extension of *us*, and what is observable to them is observable. (van Fraassen 1985, 256)

Van Fraassen's point here, I take it, is that the fact that the limits of observability can change doesn't mean that there is no line to be drawn between observable and unobservable; it just means that the line needs to be drawn *at a given time*. That much seems correct. But notice that van Fraassen is here effectively acknowledging that epistemic immigration is possible according to his own understanding of what observability amounts to.

So epistemic immigration is clearly possible on a van Fraassen-style understanding of observability, i.e. one that relativizes observability to an epistemic

5 See also Kitcher's (2001, 178) case of "Hawkeye".

community. What about epistemic upgrading? Here I wish to make another distinction, which I borrow from Paul Humphreys (2004, 4ff.), between two different kinds of epistemic upgrading. In *observational extrapolation*, one expands the range of observable values of a property whose values can already be observed within a limited range. For example, using our eyes and a ruler, we can measure the size of a given object with some particular precision—perhaps down to half a millimetre or so. Using an ordinary optical microscope, we can make the same size measurements but with considerably more precision—perhaps down to a micrometre at least (depending on which microscope we use, of course). If one *observes* the size of the object through the microscope, then this counts as a case in which one uses the microscope as a means to expand the range of observable values of the property *size*. We would thus have a case of observational extrapolation.

In the other kind of epistemic upgrading discussed by Humphreys, *observational augmentation*, one manages to increase the number of properties whose values can be observed at all. Thus, in observational augmentation, one was previously unable to observe any values of a given property, and some of those values can now be observed. If there are any cases of observational augmentation, they would presumably involve observations of properties like electron spin, ionizing radiation, and magnetic fields—presumably by means of sophisticated measurement devices of various sorts. Of course, it will no doubt be controversial whether these are really cases of observational augmentation, since many empiricists are wont to reply that when scientists claim to 'observe' the spin of an electron, or an electromagnetic field, they are using 'observe' in a loose and imprecise way that should only be taken as a manner of speaking. What scientists genuinely observe is the readings on their measurement apparatuses, the numbers on their computer screens, and so forth. So, the argument goes, scientists don't really augment their observational capacities; rather, they simply use their (unaugmented) observational capacities in a more efficient way, using instruments.

This reply may appear plausible for observational augmentation, but it is much less plausible when applied to observational extrapolation. What we observe when we use a microscope, for example, doesn't seem to be the image on the end of the microscope's eyepiece; rather, it seems more plausible that we use the microscope to observe what is placed under it. This is even clearer when it comes to instruments that we use to enhance our senses on a daily basis, such as eyeglasses and hearing aids. According to the current objection, you would strictly speaking be observing the image displayed on the inside of your eyeglasses rather than the objects in front of you as soon as you put them on your nose. If a person with normal eyesight and no glasses is standing next to you, they would of course observe the objects rather than an image on your eyeglasses. So that person would, strictly speaking, be observing something entirely different from what you are observing. At this point, one will surely start to wonder whether a notion of 'observation' that works in this way could be the notion that is most relevant to an understanding of science. Scientists do make a distinction between empirical and non-empirical results/claims, but they do not distinguish between results obtained with and without the use of eyeglasses, hearing aids, or even microscopes. In so far as constructive empiricism is supposed to make sense of science as it is actually practised, it thus seems more plausible to adopt a more flexible definition of 'observable' that allows for observational extrapolation and perhaps also observational augmentation.

A further argument against a blanket ban on the idea of observing through instruments appeals to the idea that instruments could (and have already, to a large extent) become part of 'us'—or, more specifically, of our minds. According to the *extended mind thesis* (Clark and Chalmers 1998), our minds extend out to the world in certain cases to things to which we have a sufficiently reliable access, such as notebooks and smartphones. If so, our minds can also be said to extend to certain instruments that help us perceive and process things that were previously beyond our ken. Adam Toon (2014) explores this kind of issue in a recent paper, in which he argues that this possibility "provides the realist with a new argument against the constructive empiricist view of instruments" (2014, 410) and suggests that constructive empiricism should be replaced with an "empiricism for cyborgs" in which the limits of observability depend on what sort of instruments our minds extend to at a given time.[6]

A third and final objection to the blanket ban on observation through instruments is that it seems unmotivated and arbitrary once we allow for the possibility of epistemic immigration, i. e. that the limits of observability could change due to observationally superior agents becoming part of our epistemic community. Having acknowledged that what counts as observable is relative to an epistemic community that might include, for example, humanoids with electron microscopes instead of eyes, what motivates the focus on specifically human physiology among the rest of us? If the microscope-fused humanoids count as observing what is under *their* microscopes, it seems that the rest of us (more or less normal humans) should also—on pain of drawing a completely arbitrary distinction—count as observing what is under *our* microscopes.

---

6 I think Toon is essentially correct here, although I must note that there is no such thing as "the constructive empiricist view of instruments". Constructive empiricism is perfectly compatible with any view of instruments, even if its originator, van Fraassen, has particular views about the latter. So, in my view, Toon's "empiricism for cyborgs" should be viewed as an extension of, rather than a competitor to, constructive empiricism.

Let us take stock. I have distinguished between two ways in which the limits of what is observable can be extended: In *epistemic immigration*, the relevant community is itself extended to include new, observationally superior beings. It seems undeniable that epistemic immigration is possible on a relativized notion of observability such as van Fraassen's—and indeed we find that van Fraassen explicitly acknowledges this possibility. In addition to epistemic immigration, it seems plausible that the limits of observability could be extended through epistemic upgrading, i.e. a process whereby individual members of an epistemic community increase their observational capacities. There are two kinds of epistemic upgrading, observational extrapolation and observational augmentation, and while some empiricists might be inclined to reject the possibility of observational augmentation, it seems exceedingly implausible to say that observational extrapolation would be impossible.

5 The Argument from Extended Observability

For our present purposes, the most important upshot of the considerations adduced in the last section is that the limits of what counts as observable can be extended; moreover, it is at least partly in our own power whether or not we extend these limits, since it is at least partly in our power whether we allow for epistemic immigration, and perhaps it is also in our power whether we engage in epistemic upgrading (assuming epistemic upgrading is indeed possible). We are now finally in a position to set up the argument against constructive empiricism that I am interested in exploring.

The basic idea behind the argument is that constructive empiricism implies that expanding the limits of observability carries a kind of risk for scientists that their already successful (because empirically adequate) theories will become unsuccessful (because empirically *in*adequate). It carries such a risk because once the limits of observability have expanded, all theories will be measured against a higher standard than before—viz. correctness with regard to not just all the previously observable entities but also with regard to the entities that have just become observable now that the limits of observability have expanded. This means that it is at least possible for some of the theories that were previously empirically adequate—given the previous limits for observability—to become empirically inadequate—given the new limits for observability. So constructive empiricism implies that expanding the limits of observability carries a risk of going against the aim of science, i.e. of impeding scientific success. Thus, all other things being equal, constructive empiricism implies that scientists have an important scientific reason not to seek to expand the limits of observability.

To illustrate, suppose that a given community of scientists were offered the choice of accepting into their community a number of beings who could perceive electromagnetic fields in much the same way that we see ordinary light. (This would be a case of epistemic immigration.) Or, to use a different kind of example, suppose a new instrument could be constructed through which magnetic fields could be detected with the same reliability and effortlessness with which most humans currently perceive ordinary light. (This would be a case of epistemic upgrading.) From the point of view of a scientific realist, there is no reason whatsoever not to allow this type of epistemic immigration or upgrading, since the only possible upshot is that we obtain another means to (reliably) measure electromagnetism, which in turn would increase our chances of obtaining true theories about electromagnetism and related phenomena.

Things will look differently from constructive empiricism's point of view. This new means of observing electromagnetism might help us get better—more empirically accurate—theories in various ways. But it will also elevate the standards against which any given theory is judged, and thereby make those standards harder to achieve. In particular, a theory of electromagnetism that had been empirically adequate before—because correct about all the entities that were observable at the time—could become empirically inadequate afterwards—because incorrect about some new set of entities that became observable by virtue of the epistemic immigration or upgrading that would just have taken place. So constructive empiricism implies, in a way scientific realism does not, that epistemic immigration and upgrading involves risking that currently successful theories would become unsuccessful.

It seems to me that this implication of constructive empiricism is problematic for two related reasons. First, if there was a risk for science to regress in this way through epistemic immigration and upgrading, we should expect scientists and scientific policy makers to be at least somewhat aware of this risk—if not explicitly in their pronouncements and stated aims, then at least implicitly through their planning and actions. However, I think it's fair to say that there is no indication from scientific practice that scientists and scientific policy makers are in any way cautious or wary with regard to epistemic upgrading and immigration. If they were, we should expect a lot more energy and resources to be spent on researching where exactly the current boundaries of observability are located—exactly which entities are observable and which entities are not—and on coming up with plans for how to avoid transgressing those boundaries. The fact that we don't see anything of this sort in scientific practice is thus a strike against constructive empiricism as an account of the aim of science.

A second, but related, reason it is problematic for constructive empiricism to imply that epistemic immigration and upgrading involves risking that successful theories would become unsuccessful concerns its normative implications. An often-overlooked fact about the scientific realism debate is that the various positions in this debate have important normative implications.[7] For example, a scientific realist—who values truth beyond empirical adequacy—thinks that the work of a scientist is not finished once they have obtained empirically adequate theories of a given phenomenon; rather, the scientists must also choose which of a set of empirically adequate theories is also true. So while a scientist who follows constructive empiricism's advice would move on to researching other phenomena as soon as they are convinced that empirical adequacy has been achieved, the realist-inclined scientist might rationally choose to spend their time and resources on finding the true theory of a phenomenon for which they have already achieved empirical adequacy.

The point here is that whether scientists follow constructive empiricism or scientific realism will have real-life normative implications for how they conduct their research. Whether we, as philosophers, prefer constructive empiricism or scientific realism (or some other account of the aim of science) should thus depend—at least in part—on whether we agree with each account's normative implications. The situation here is analogous to that in normative ethics, where we choose between ethical theories, e. g. utilitarianism and deontology, at least partly by appealing to the normative implications of each theory. For example, it is a *prima facie* objection to utilitarianism that it seems to recommend involuntary organ donations in which a single person is sacrificed to save the lives of five other people; similarly, it is a *prima facie* objection to Kantian deontology that it offers a blanket ban on lying, even to axe-murderers who are searching for their next victims.

What normative recommendation would constructive empiricism give a community of scientists who are faced with a choice of either (a) allowing or banning epistemic immigration, e. g. from creatures who are able to directly perceive electromagnetic fields, or (b) engaging in or abstaining from epistemic upgrading, e. g. by inventing and constructing a highly reliable and accessible instrument for perceiving electromagnetic fields? Of course, the answer will depend on exactly what other foreseeable benefits and drawbacks there are to each course of action. But one important consideration for the constructive empiricist would be that extending the limits of observability in this way risks making empirically successful theories unsuccessful. So there will be circumstances—however contrived and convoluted—in which constructive empiricism implies that epistemic immigration and upgrading should be avoided or even banned (at least in so far as we are seeking scientific success). By contrast, scientific realism will have no such implications,

7 For more on this general point, see Dellsén (2019).

since for the realist there is never any risk of changing which theories count as successful. After all, what counts as *the truth* is always and everywhere the same, regardless of what is observable within one's epistemic community.

Now, isn't there something troubling, even absurd, in suggesting that epistemic immigration and upgrading could be something to be avoided? In both cases, we are simply extending the realm of entities that can be observed by us. How could that possibly be a bad thing? From an empiricist perspective, i.e. from the perspective of those who prioritize the empirical over the non-empirical aspects of science, epistemic immigration and upgrading give us a world in which a larger proportion of entities are located on the favoured side of the empirical/non-empirical divide. Again, it is hard to see any reason to think this could ever be a bad thing. So the implication from constructive empiricism that it would (at least sometimes) be unadvisable for scientists to allow epistemic immigration or engage in epistemic upgrading appears to be an awkward normative implication, much like utilitarianism's (apparent) implication regarding involuntary organ donation and Kantian deontology's (apparent) implication regarding a blanket ban on lying.

6 Two Empiricist Rejoinders (with Replies)

How might a constructive empiricist, such as van Fraassen himself, respond to the argument of the previous section? Before concluding this paper, I will consider two possible responses that I think have a great deal of *prima facie* plausibility—even though I don't think either is ultimately successful. According to the first response, any change to the limits of observability is, by constructive empiricism's lights, a change in the criterion of success in science, and thus a change in the nature of science itself. As a result, comparing scientific success before and after a change in observability is effectively a comparison of success in different enterprises—an earlier science and a later science—for which there is no common standard of comparison. And since there is no common standard of comparison, any such comparison of scientific success before and after a change in observability is misguided.

To develop this response a little further, consider van Fraassen's analogy between science and chess. In explaining what he means by attributing an *aim* to science, van Fraassen notes that he means this in the same sense in which "[t]he aim of the game of chess is to checkmate your opponent" (van Fraassen 1980, 8). Now, suppose the rules concerning *checkmate* are changed so that what we currently refer to as a stalemate is henceforth considered as a kind of checkmate. Thus the 'aim' of the game of chess—its criterion of success—would now be different from what it was before, and chess players would have to adapt to the new rules, e.g. by making appropriate changes in their training and strategies. Now we can ask ourselves the following question: Would the game that is being played after the rule change be the same game as was being played before? Or would we rather be playing a different, albeit very similar, game afterwards? If we say that we are playing a different game after the rule change, then it seems that success and failure before and after the change cannot really be compared in a meaningful way. That would be to compare success and failure in one game with success and failure in another, e.g. like comparing success in tennis with success in badminton. Such across-game comparisons do not seem to make any sense.

Similarly, a constructive empiricist might argue that if the limits of observability change from one time, t_1, to another, t_2, we in effect have two separate criteria for success at t_1 and t_2 according to constructive empiricism. These can be roughly stated as follows:

> *Empirical adequacy$_1$:* Being correct about what is observable for the relevant scientific community at t_1.
>
> *Empirical adequacy$_2$:* Being correct about what is observable for the relevant scientific community at t_2.

Now, since the basic criterion of success in an endeavour such as science is arguably constitutive of it, this would mean that we really have two distinct (but of course very similar) endeavours at t_1 and t_2. These two endeavours can be labelled *science$_1$* and *science$_2$*. Of course, since science$_1$ and science$_2$ are so similar, one would be forgiven for confusing the two in most contexts (just as someone who confuses regular chess with stalemate-inclusive chess would surely be forgiven that in most contexts). Nevertheless, the two endeavours would be genuinely distinct, as evidenced by the fact that their constitutive criteria of success would be different.

Now the point of all of this would be to show that, in case of epistemic immigration or upgrading, there wouldn't strictly speaking be a single endeavour, science, such that success and failure in that endeavour can be meaningfully compared at t_1 and t_2 respectively. Of course, we can still ask whether the endeavour we have is successful according to each of the criteria operative at t_1 and t_2 respectively, i.e. empirical adequacy$_1$ and empirical adequacy$_2$. But changing the limits of observability will not itself have any effect on whether our theories count as empirically adequate$_1$, for example: If they were empirically adequate$_1$ before the change, they will be empirically adequate$_1$ afterwards as well (assuming all else remains equal in the process). And the same will be true, *mutatis mutandis*, for empirical adequacy$_2$. But if it's misguided in this way to compare scientific success and failure before and after epistemic immigration or upgrading, i.e. success and failure in science$_1$ and science$_2$ respectively, then the argument I presented in the previous section doesn't seem to get off the ground.

So that's the empiricist rejoinder to the argument of the last section. I have said that I do not ultimately think it is successful. So what's my reply? Well, consider the issue from the point of view of scientists themselves. For scientists, what matters is to satisfy the criterion of scientific success, whatever it is, that is in play at for them at a given time. It does not matter to working scientists whether they are, strictly speaking, doing science$_1$ or science$_2$; what matters is only that they be successful at whatever enterprise they are engaged with. So, in their deliberation about what they should do in order to satisfy the criterion of success, scientists should take into account what the criterion will be in the near-to-distant future when their action takes place or bears fruit, as opposed to what the criterion is at the time of deliberation. And this holds even if the criterion at that point will be a criterion in a different endeavour, science$_2$, since they will then have gone over from doing science$_1$ to doing science$_2$.

To understand this point fully, it may be useful to go back to the chess analogy. A chess player who knows or strongly suspects that the rules of chess will change soon in the way we envisioned before (making a stalemate count as a kind of checkmate) should take that into account in her deliberation, e.g. in deciding how to prepare for her next tournament. It doesn't matter to the chess player whether the game that is being played at the next tournament is, strictly speaking, a different game than what she is used to play (chess$_2$ as opposed to chess$_1$, if you will). The chess player simply wants to be successful relative to whatever criterion of success is in play at the time at which the game is played. Similarly, I am suggesting, a scientist will want to satisfy the criterion of success for the enterprise in which she is working—regardless of whether that criterion is empirical adequacy$_1$ in science$_1$ or empirical adequacy$_2$ in science$_2$.

Accordingly, it seems to me that if it's in scientists' power to prevent the criterion of success in their discipline (whatever it is)[8] from changing between t_1 and t_2 in a way that makes their own theories unsuccessful according to the criterion in play at t_2, then they would be rational in doing so—at least in so far as they seek scientific success in the first place. And now we are back where we started: If constructive empiricism is correct that the criterion for success in science is empirical adequacy, then scientists who expand the limits of what can be observed risk making theories that were successful according to the previous criterion for

8 The parenthesis here is meant to indicate that "criterion of success" is being used as non-rigid designator here to whatever is the criterion in the discipline that scientists happen to be engaged in.

success into scientifically unsuccessful theories according to the criterion in place after the expansion of the limits of observability. This holds even if criteria of success are constitutive of enterprises like science, because then these scientists can be described as changing the enterprise in which they are engaged from one in which their theories are successful—what I've been calling science_1—into one in which their theories are unsuccessful—science_2. As before, however, it doesn't seem as if there is any indication in scientific practice that scientists are concerned about 'risks' of this kind, nor does it seem to be a reasonable normative implication that it would sometimes be irrational for conscientious scientists to avoid or ban epistemic immigration and upgrading.

A second empiricist rejoinder is to concede that the argument of the previous section does undermine constructive empiricism on the original formulation of the view, but insist that the argument can be avoided by making a small modification to constructive empiricism. Specifically, the constructive empiricist could insist that the aim of science is not just for one's theories to be empirically adequate, but also to increase the range of observable entities. So, on this picture, science has two related aims: The first is to have theories that are correct about observables; the second is to increase the number of entities that count as observable. Given this additional aim of science, the constructive empiricist (of this modified variety) can explain why scientists ought to expand the limits of observable by appealing to the fact that doing so makes additional things observable, thus satisfying the second aim of science.[9]

The first thing to say about this rejoinder is that it is clearly somewhat *ad hoc*, in being a modification to a theory that is specifically designed to avoid a problem and isn't independently motivated. Still, *ad hoc* theories can be true (as the history of science surely teaches us) so that by itself is not a knock-down argument against this rejoinder. A related point is that this conjunctive version of constructive empiricism would clearly be *less simple* than scientific realism, which posits only one aim for science, viz. that its theories be true). To this van Fraassen (or a van Fraassenite) could reply that he (she) places no epistemic weight on simplicity in scientific theory choice, so why should they be concerned with having a less simple account of science than the realist? So, again, I don't think pointing to the complexity of this new empiricist account of the aim of science constitutes a knock-down argument either.

But there is a more serious problem in the vicinity here. That problem emerges when we consider the issue from the point of view of a scientific realist. From the realist's perspective, the second aim proposed by the current rejoinder is per-

9 Many thanks to Hedda Hassel Mørch and Andreas Brekke Carlsson for raising this concern.

fectly explicable as a means to obtaining true theories. Scientists should aim to expand the limits of observability, i. e. turn unobservable entities into observable ones, because science aims to discover the truth about *all* entities—not just the observable ones—and expanding the limits of observability enables us to confirm theories about previously-unobservable entities more strongly and more efficiently. The scientific realist can thus *explain* why the constructive empiricist must add the second aim in order to avoid the present problem. So, in a sense, adding this second aim to constructive empiricism is not *ad hoc* from the scientific realist's perspective.

It seems to me that this last point—that scientific realism can explain why an apparently *ad hoc* modification must be made to constructive empiricism, while constructive empiricism (in its original or modified variety) clearly cannot—is the most serious problem with the second empiricist rejoinder. In science and philosophy, one must sometimes learn to live with theories that are not as simple as one would like, or that have been modified *ad hoc* to avoid some problems; but it's another thing entirely to cling to a theory whose *ad hoc* modification is unmotivated by its own lights and yet perfectly explicable according to a rival theory.

7 Conclusion

I have argued that there is a tension in constructive empiricism's view of science, coupled with van Fraassen's relativized notion of observability. Specifically, I have argued that constructive empiricism implies that scientists have reason to prevent or discourage expanding the limits of what could be observed, because such an expansion risks making empirically adequate theories into empirically inadequate ones. This implication, I argued, is problematic—in part because there doesn't seem to be anything in scientific practice to suggest that scientists themselves take there to be any risk of this kind involved in expanding the limits of observability, and in part because this has the awkward normative implication that scientists should sometimes abstain from or even ban certain ways of expanding the current limits of observability.

I am not going to seriously propose a solution to the problem, but I'll end with a question that I think is worth thinking about. Might it be that this problem can be avoided by defining empirical adequacy not in terms of what is observable to the community at a time, as van Fraassen does, but rather in terms of what *has been, is, or will be* observable to the community—i. e. in terms of what is observable not at a specific time (now or in the past) but at *any time whatsoever.* I am not entirely sure this solves all the problems of the kind I've been concerned with, and it also might bring new problems that I haven't thought of. But it might be a step in the right direction for those who, like me, are intrigued and impressed by the view of science we find in van Fraassen's constructive empiricism.

Acknowledgement: I am grateful for helpful feedback from participants at the Lauener Symposium at the University of Bern in September 2018 and my colleagues at Inland Norway University of Applied Sciences. I would like to especially thank Bas van Fraassen for inspiration and discussions in Bern.

References

Alspector-Kelly, M. 2001. Should the Empiricist Be a Constructive Empiricist? *Philosophy of Science* 68: 413–433.

Asay, J. and Bordner, S. 2015. A Modest Defense of Manifestationalism. *Synthese* 192: 147–161.

Churchland, P. M. 1985. The Ontological Status of Observables: In Praise of the Superempirical Virtues. In: *Images of Science: Essays on Realism and Empiricism, with a Reply from Bas C. van Fraassen*, eds. P. M. Churchland and C. A. Hooker, 35–47. Chicago: University of Chicago Press.

Clark, A. and Chalmers, D. 1998. The Extended Mind. *Analysis* 58: 7–19.

Dellsén, F. 2017. Reconstructed Empiricism. *Acta Analytica* 32: 95–113.

Dellsén, F. 2019. Should Scientific Realists Embrace Theoretical Conservatism? *Studies in History and Philosophy of Science* 76: 30–38.

Dicken, P. 2009. On the Syntax and Semantics of Observability: A Reply to Muller and van Fraassen. *Analysis* 69: 38–42.

Dicken, P. and Lipton, P. 2006. What Can Bas Believe? Musgrave and van Fraassen on Observability. *Analysis* 66: 226–233.

Geach, P. T. 1969. *God and the Soul*. London: Routledge.

Humphreys, P. 2004. *Extending Ourselves: Computational Science, Empiricism, and Scientific Method*. Oxford: Oxford University Press.

Kitcher, P. 2001. Real Realism: The Galilean Strategy. *The Philosophical Review* 110: 151–197.

Monton, B. and van Fraassen, B. C. 2003. Constructive Empiricism and Modal Nominalism. *British Journal for the Philosophy of Science* 54: 405–422.

Muller, F. A. and van Fraassen, B. C. 2008. How to Talk about Unobservables. *Analysis* 68(3):197–205.

Musgrave, A. 1985. Constructive Empiricism and Realism. In: *Images of Science: Essays on Realism and Empiricism, with a Reply from Bas C. van Fraassen*, eds. P. M. Churchland and C. A. Hooker, 245–308. Chicago: University of Chicago Press.

Psillos, S. 1996. On van Fraassen's Critique of Abductive Reasoning. *The Philosophical Quarterly* 46: 31–47.

Railton, P. 1989. Explanation and Metaphysical Controversy. In: *Scientific Explanation*, eds. P. Kitcher and W. Salmon, 13–220. Minnesota: University of Minnesota Press.

Rosen, G. 1994. What Is Constructive Empiricism? *Philosophical Studies* 74: 143–178.

Toon, A. 2014. Empiricism for Cyborgs. *Philosophical Issues* 24: 409–425.

van Fraassen, B. C. 1980. *The Scientific Image*. Oxford: Oxford University Press.

van Fraassen, B. C. 1985. Empiricism in the Philosophy of Science. In: *Images of Science: Essays on Realism and Empiricism, with a Reply from Bas C. van Fraassen*, eds. P. M. Churchland and C. A. Hooker, 245–308. Chicago: University of Chicago Press.

van Fraassen, B. C. 1994. Gideon Rosen on Constructive Empiricism. *Philosophical Studies* 74: 179–192.
van Fraassen, B. C. 2001. Constructive Empiricism Now. *Philosophical Studies* 106: 151–170.
van Fraassen, B. C. 2007. From a View of Science to a New Empiricism. In: *Images of Empiricism: Essays on Science and Stances, with a Reply from Bas C. van Fraassen*, ed. B. Monton, 337–382. Oxford: Oxford University Press.
Williamson, T. 1992. Vagueness and Ignorance. *Proceedings of the Aristotelian Society*, Supplementary Volume 66: 145–162.

Milena Ivanova

Aesthetic Virtues and Theory Acceptance

Abstract: Trusting a theory involves considerations that go beyond logic and experience; scientists often appeal to aesthetic criteria, such as the theory's elegance, unity, coherence, simplicity. But what role exactly do these criteria play in our choices: do they aid us to make the most convenient choices or do they lead us to the truth? In this chapter I discuss several strategies developed by scientific realists to argue in favour of an epistemic understanding of theory virtues as truth-tracking criteria and argue against them. Drawing insight from van Fraassen's focus on agency in theory choice, I will offer further support for the pragmatic understanding of theory virtues.

1 Introduction

Trusting a theory involves several considerations that can go beyond the empirical adequacy of the theory. When scientists form their epistemic attitudes towards a theory, they look for empirical and logical considerations: is the theory making successful predictions; does it predict beyond its intended domain; is it internally consistent? But one's attitude towards a theory is also formed by considerations that go beyond logic and experience, by considering extra-empirical, often also called aesthetic, criteria such as the theory's elegance, unity, coherence, simplicity. These virtues of the theory often dictate theory choice, especially when the data is not available to conclusively support one theory. But what role exactly do these virtues play in our choices? Do they help us choose the most convenient theory, or can we claim that a theory that satisfies a list of such aesthetic properties is more likely to be true? In this chapter I discuss several strategies developed by scientific realists to argue in favour of an epistemic understanding of theory virtues. These arguments aim to support the idea that there is a special epistemic relationship between the virtues of a theory and its likelihood of being true. Instrumentalists, positivists and constructive empiricists have long opposed this idea, with Bas van Fraassen presenting a challenge to the epistemic take on virtues. I will discuss three strategies that have been developed to defend the epistemic significance of theory virtues, present some problems for these accounts, and return to van Fraassen's focus on agency and pragmatics in theory choice, offering further support to the pragmatic understanding of theory virtues.

https://doi.org/10.1515/9783111019802-010

2 Theory Virtues as Pragmatic Considerations

It is well known that scientists praise theories for their aesthetic appeal, especially when the evidence is insufficient or unavailable to support them. Many express the belief that a simple, beautiful, elegant, unifying theory is more likely to be true. For instance, Paul Dirac and Werner Heisenberg were among the physicists who claimed that trust in the theory of relativity was justified even prior to its empirical confirmation because the theory was too beautiful not to be true. Specifically, Dirac argues

> What makes the theory of relativity so acceptable to physicists is its *mathematical beauty*. One has a great confidence in the theory arising from its great beauty, quite independent of its detailed successes. (Dirac 1980, 40)

Similarly, Heisenberg claims "if nature leads us to mathematical forms of great simplicity and beauty we cannot help thinking that they are 'true', that they reveal a genuine feature of nature" (Heisenberg 1971, 68). The Nobel laureate Subrahmanyan Chandrasekhar further defends the idea that believing in the truth of relativity theory is motivated by its overwhelming aesthetic quality. The general theory of relativity, Chandrasekhar argues, achieved the unification of our fundamental concepts of space and time, matter and motion with "unerring sense for mathematical elegance and simplicity" (Chandrasekhar 1987, 71). The Nobel Laureate Murray Gell-Mann, whose search for symmetry led to great discoveries in particle physics and the advancement of the standard model, has similarly argued that beautiful and elegant theories in fundamental physics are more likely to be correct (Hossenfelder 2018, 24). And when it comes to confidence in string theory, for which we have no empirical confirmation, the theoretical physicist John Schwarz argues that "the mathematical structure of string theory was so beautiful and has so many miraculous properties that it had to be pointing towards something deep" (quoted in Green 1999, 137).

Contrary to the above claims, in the *Scientific Image* Bas van Fraassen challenged the idea that the virtues of a theory can be anything but pragmatic. For van Fraassen, the aim of science should be seen as aiming at empirical adequacy. When a theory is accepted, this acceptance involves only belief in the empirical adequacy of the theory, contrary to scientific realism, which involves belief in empirical adequacy plus additional commitments towards the existence and properties of unobservable entities corresponding to the theoretical terms of scientific theories. What the additional content scientific realists commit to is a matter for debate, with realists often looking into cases of theory change to identify continuity at the level of structure of essential parts of theories that are responsible for its success. While agreeing that when we praise theories, we often need to appeal to non-empirical factors, like the elegance, simplicity and scope of the theory, for van Fraassen these considerations are purely "human concerns, a function of our interests and pleasures, which make some theories more valuable or appealing to us than others. These values provide reasons for using a theory but cannot rationally guide our epistemic attitudes and decisions" (1980, 87). The virtues of a theory might be useful in the development and acceptance of theories that are empirically successful, but they should not be taken to indicate the truth of the theory and the claims it makes about the unobservable aspects of the world.

The pragmatic reading of theory virtues was also defended by early twentieth-century empiricists like Ernst Mach, Henri Poincaré and Pierre Duhem. In the *Science of Mechanics*, Mach argues that we should always aim to explain the phenomena in the most conceptually economical manner, because simple theories are more convenient for us. He argues that science should be seen as an economical description of our observations, with simplicity being a guiding principle in the construction and evaluation of scientific hypotheses. According to Mach, "[t]he aim of science is to be an 'economy of thought', to offer us a simple and concise classification of the observable appearances and enable the prediction of phenomena" (Mach 1984, 577). Simplification or economy is thus taken to be the ultimate aim and virtue of scientific theories.

Similarly, Poincaré claims:

> [T]he object of mathematical theories is not to reveal to us the real nature of things; that would be an unreasonable claim. Their only object is to coordinate the physical laws with which physical experiments make us acquainted, the enunciation of which, without the aid of mathematics, we should be unable to effect. (Poincaré 2001 [1902, 1905, 1908], 155)

His attitude towards theoretical virtues like simplicity, unity and scope was that they are important insofar as they are linked to utility, in that a simpler theory with broad scope is more convenient to use and aids our understanding, making these theory virtues instrumental as guiding principles in the construction and choice of theories. Importantly, Poincaré argues that theories that exemplify these virtues are considered beautiful, pleasing a scientists' aesthetic sensibility and generating a sense of pleasure for the mind, which seeks order in the phenomena.[1] According to him,

---

1 Note that often in philosophy of science properties of theories such as simplicity, scope and elegance are referred to as theory virtues, rather than as aesthetic properties of theories, due to the worry that the latter connote subjectivity. However, some philosophers have taken the aesthetic aspect of scientific products seriously and offered solutions to the subjectivity claim. This move

> care for the beautiful leads us to the same selection [of theories] as care for the useful. Similarly economy of thought, that economy of effort which, according to Mach, is the constant tendency of science, is a source of beauty as well as a practical advantage. (Poincaré 2001 [1902, 1905, 1908], 369)

In the *Aim and Structure of Physical Theories* Duhem dedicates a lot of attention to the problem of theory choice and the non-empirical values guiding scientists' decisions. Similar to Mach and Poincaré, Duhem claims that the aim of science is not to uncover "essences" behind the observable appearances, but to classify observational laws in a concise way. As he argues, physical theory is "a system of mathematical propositions, deduced from a small number of principles, which aim to represent as simply, as completely, and as exactly as possible, a set of experimental laws" (Duhem 1954, 19). The employment of theory virtues, for Duhem, was inevitable in scientific practice due to the problem of underdetermination of theory by the evidence. Given that multiple different theories can be consistent with the same set of data, a choice is often needed between alternative theories that cannot be grounded either in the laws of logic nor in experience. Thus, he claims, there is no "absolute principle" to dictate a decision. Duhem's views on theory choice have caused some interpretative disagreement, especially because he claims that when scientists disagree about the merits of a theory, they will choose on the basis of their intuitions, or what he calls "good sense".[2] Duhem certainly acknowledges the pragmatic role of theory virtues when he argues that theory choice is guided by considerations of "elegance, simplicity, and convenience" (Duhem 1954, 288). Importantly for him theory choice based solely on such considerations does not guarantee convergence on a choice, since the virtues taken into consideration are contingent upon the schools and practitioners that employ them, an argument later developed by Thomas Kuhn and others. Ultimately, for Duhem, empirical confirmation is the only source for conclusive decision between two competing theories.

While the above arguments defend the use of theory virtues but only to the extent that they are seen as pragmatic considerations, reflecting agents' aims and preferences, many believe that theory virtues are not just pragmatic but epistemic, such that an aesthetically virtuous theory is more likely to be true than a

---

is motivated by the desire to account for the fact that the language scientists use when praising scientific theories is importantly similar to the language used to praise works of art (Breitenbach 2013; McAllister 1996; Kosso 2002). Furthermore, there is ground to infer that a theory with such virtues generates an aesthetic response in the scientists, similar to the response they have when appreciating works of art (Zeki et al. 2014). For a recent engagement in the aesthetics of science, see Bueno et. al (2017), French and Ivanova (2020), Ivanova (2017a), Ivanova and Murphy (2023) and Ivanova (2023).

2 For a discussion of Duhem's notion, see Stump (2007), Ivanova (2010, 2021).

complex, inelegant, messy or "ugly" one. To defend this idea, scientific realists have devised different arguments to show that the virtues of the theory have epistemic significance. In the next section I examine three such arguments: the ontological argument; the track record argument; and argument from convergence in practice.

3 Defending the Epistemic Import of Theory Virtues

First is the ontological argument, often assumed by many practising scientists, including Isaac Newton, Paul Dirac and Subrahmanyan Chandrasekhar. The idea is that a simple, elegant, unified or beautiful theory is more likely to be true simply because it reflects the simplicity, elegance, unity or beauty of nature. Taking simplicity as an example, this argument is advanced by Newton in the *Philosophiæ Naturalis Principia Mathematica* where he takes it as a methodological rule that we need to formulate the simplest laws because "nature is simple and does not indulge in the luxury of superfluous causes" (Newton 1999 [1687], 794). The justification for the principle of economy is that nature itself must be simple. A similar argument is developed by Richard Swinburne, who takes simplicity to be evidence for a theory's truth because it reflects the simplicity of nature. In *Simplicity as Evidence for Truth*, Swinburne argues that

> The simplest hypothesis proposed as an explanation of phenomena is more likely to be the true one than is any other available hypothesis, it is an ultimate *a priori* epistemic principle that simplicity is evidence for truth. (Swinburne 1997, 1)

These arguments are taken to be circular: our confidence in the claim that nature is simple, unified or beautiful is justified upon reflection on our current best theories, which in their turn are meant to be true in light of their reflecting the beauty, simplicity or unity in the world. It is also problematic to assume the simplicity or unity of nature a priori, since reflecting upon some of our contemporary theories can also lead us to make the very contrary conclusion. For instance, if we take the standard model of particle physics, often claimed to be an inelegant and even "ugly" theory, to reflect the nature of reality, we would perhaps have to conclude that the world is not simple or unified as we would like it to be, and perhaps there is more complexity and irregularity in the world.

In the *Scientific Image* van Fraassen argues convincingly against the ontological argument, claiming that

> It is surely absurd to think that the world is more likely to be simple than complicated (unless one has certain metaphysical or theological views not usually accepted as legitimate factors in scientific inference). The point is that the virtue, or patchwork of virtues, indicated by the theory is a factor in theory appraisal, but does not indicate special features that make a theory more likely to be true. (van Fraassen 1980, 90)

And targeting specifically the arguments defending the link between simplicity and truth, van Fraassen argues that "we have no evidence that reality is simple. It is therefore mistaken to believe that simple theories are more likely to be true than complex ones" (van Fraassen 1980, 90).

Departing from such a priori arguments, many scientific realists have advanced sophisticated empirically motivated defences of the relationship between theory virtues and truth. One such defence is based on the historical track record of successful and virtuous theories of the past. The aim of these arguments is to establish that it is an empirical fact that theoretical virtues lead to (approximately) true theories, claiming that the virtues of a theory give indirect evidence for its truth. The argument builds on the idea that we can reflect on the theory virtues exemplified by mature past theories and infer that a theory is likely to be true if it possesses the virtues with good track record. Richard Boyd (1984) for instance argues that theories possessing theoretical virtues that have been instantiated by previously successful theories, such as simplicity, elegance and unity, will be ranked as more plausible. A reflection on the past successful theories that satisfy the scientific realist's criteria of commitment (for instance theories with novel predictive power and maturity) can justify us to infer, for instance, that a unifying theory is more plausible, since our best theories in physics achieved high levels of unification. Copernicus unified terrestrial and celestial phenomena in one planetary framework, Maxwell unified electric and magnetic phenomena, Einstein unified gravity and inertial mass, and the concepts of space and time into a spacetime continuum. This reflection serves to boost our confidence in the unifying power of theories, making them more plausible when there is as yet insufficient empirical evidence to support them.

Wesley Salmon goes further in showing how these virtues can bear upon the confirmation of a theory by arguing that the past record of success will ultimately determine the prior probability of a theory. In his "The Appraisal of Theories: Kuhn Meets Bayes" Salmon reconciles Kuhn's idea that certain theory virtues guide the context of discovery with Bayes' theorem, which provides the conditions of confirmation of a theory. The starting point is Kuhn's discussions of theory virtues, especially his 1977 paper where he takes a set of theory virtues to be praised by scientists independently of background commitments or paradigm. Kuhn famously argues, as had Duhem before him, that theory choice cannot be taken to be a matter of proof, a conjunction only of logical and empirical considerations, but rather also involves judgement and persuasion. However, scientists would agree on the virtues of a good theory, with the classic list of virtues providing a basis for objective theory comparison: "these five criteria—accuracy, consistency, scope, simplicity, and fruitfulness—are all standard criteria for evaluating the adequacy of a theory" (Kuhn 1977, 322). Salmon takes Kuhn's point and argues that by using Bayes' theorem, one can show how plausibility arguments can play a role not only in the context of discovery but also in the context of justification. He argues that

> [a]s long as the [hypothetico-deductive] schema is used to characterise scientific confirmation, plausibility arguments are related to heuristics as an aid in the generation of interesting and promising hypotheses. Bayes' theorem, in contrast, shows that they play an indispensable role in the appraisal of theories. (Salmon 1990, 327)

Boyd's and Salmon's accounts are endorsed by Stathis Psillos, who also explains how the past record of the theory virtues of successful theories can determine the plausibility and prior probability of a theory, explaining that

> [t]he past record of mature scientific theories can be seen as the background knowledge in the light of which the plausibility of emergent scientific theories can be judged and estimated [...] if theories which have not been subjected to ad hoc adjustments have tended to be better supported by the evidence than theories with ad hoc features, then this consideration should be used in assessing the prior probability of other theories, in order to rank higher theories with no ad hoc features. (Psillos 1999, 172)

Of course the plausibility of this account depends on us identifying what virtues have been exemplified in our successful theories and providing conditions for when a theory is instantiating a certain virtue. As Psillos himself acknowledges,

> Naturally, finding out which theoretical virtues have been associated with well-confirmed theories can be the outcome only of substantive empirical-historical research. (Psillos 1999, 172)

Not many realists have engaged with the investigations of case studies with the aim of showing what kind of theory virtues have been linked to success, with the work of James McAllister (1996) and Sam Schindler (2018) being an exception. But even if we grant this argument, it is difficult not to question the idea that the list of the aesthetic qualities linked to success are anything but correlate with empirical success rather than truth.

My concern is that these track record arguments don't have the strength to push in the realist direction for two reasons. First, plenty of successful and virtuous theories have been found to be false, despite satisfying all the criteria for commitment of the scientific realist, such as novel predictive power and maturity. Second, we can make indicative projections from theories that lack virtues (commonly takes as "ugly" or inelegant) to success. The problem is that there are many aesthetically pleasing theories that have turned out to be false and many aesthetically displeasing theories that fit with the highest requirements for commitment of the realist that would justify opposite inferences. On the one hand, we have the virtuous theories that failed. Copernicus' heliocentric system was abandoned in favour of Kepler's despite the initial negative reaction to the idea of abandoning a system based on perfect cycles to ellipses, considered at the time aesthetically inferior. We consider Newton's gravitational theory at best as an approximation, but technically-speaking false. Aristotelian biology certainly satisfies a number of virtues such as simplicity and scope, but is false. Darwinian evolution lacks the clear cut boundaries between species and adds chance in their evolution, which is much messier than Aristotelian appeal to telos and essence and proper function. And when it comes to theoretical projects in high-level physics, unificationist projects have been abandoned due to difficulties despite exemplifying the virtues of good theory. The Kaluza-Klein theory, for instance, provided an elegant way to unify gravity and all the other gauge fields but was abandoned despite it being regarded highly due to its elegance and simplicity. Sabine Hossenfelder's (2018) *Lost in Math: How Beauty Leads Physics Astray* is full of illuminating examples from theoretical physics where theories of great aesthetic merit are abandoned due to their lack of empirical success. Hossenfelder questions whether aesthetic values such as simplicity and naturalness can be guides in science and argues that we should be wary of relying on beauty to guide our belief in theories, and that beauty can often be a systematic bias that leads away from developing successful theories. Hossenfelder rightly points out that

> The modern faith in beauty's guidance is, therefore, built on its use in the development of the standard model and general relativity; it is commonly rationalized as an experience value: they noticed it works, and it seems only prudent to continue using it. so it seems reasonable to keep using it. (Hossenfelder 2018, 26)

This track record reasoning certainly captures how many scientists reason (whether implicitly or explicitly), but reflecting on the history of science should make us wary of such a link. Hossenfelder notes that "not only does the history of science thrive with beautiful theories that turned out to be wrong, but on the flipside we have the ugly ideas that turned out to be correct" (Hossenfelder 2018, 31).

Many of the theories that satisfy some of the realist's main criteria of commitment, such as maturity or novel predictive power, are not theories praised for their virtues, in fact they are often explicitly deemed to be "ugly", yet very successful.

When it comes to placing confidence in the track record of "ugly" theories, our current best theories in physics—e.g. the standard model and quantum mechanics—provide examples of great empirically successful theories, ticking the list of the realist requirements for commitment, but are rarely used as examples of virtuous or beautiful theories, quite the contrary. As Kaku and Thompson argue in the context of the standard model

> [A]t present, there has been no experimental deviation from the Standard Model. Thus, it is perhaps the most successful theory ever proposed in the history of science. However, most physicists find the Standard Model unappealing because it is exceptionally ugly and asymmetrical. [...] The reason why the Standard Model is so ugly is that it is obtained by gluing, by brute force, the current theories of the electromagnetic force, the weak force, and the strong force into one theory. (Kaku and Thompson 1997, 75)

For both these reasons I believe the realist attempt to show the link between approximate truth and theory virtues is difficult to establish since we can find plenty of examples to base an inductive argument on, some in favour of the link and some against it, making this strategy inconclusive.

A further challenge to the idea that theory virtues are epistemic comes from the fact that these virtues might not be stable or fixed. Beautiful or lovely theories are taken to be elegant, simple, unifying and broad in scope, but what theory satisfies those criteria certainly varies not only between fields but also in the very same field between different communities and different times. This problem is noted by Duhem, who argues that theory virtues are

> essentially subjective, contingent, and variable with time, with schools, and with persons; as serious as these motives may be in certain cases, they will never be of a nature that necessitates adhering to one of the two theories and rejecting the other, for only the discovery of a fact that would be represented by one of the theories, and not by the other, would result in a forced opinion. (Duhem 1954, 288)

Thomas Kuhn further develops this argument in the *Structure of Scientific Revolutions* where he emphasizes the idea that paradigms impose their own values and "each paradigm will be shown to satisfy more or less the criteria that it dictates for itself and to fall short of a few of those dictated by its opponent" (Kuhn 1996 [1962], 109). When revisiting the problem of theory choice in *The Essential Tension*, in the widely discussed article "Objectivity, Value Judgment, and Theory Choice" mentioned above, Kuhn identifies five criteria in theory choice that are "fixed once and for all, [...] unaffected by their participation in transitions from one theory to another': empirical accuracy, consistency, scope, simplicity, and fertility" (Kuhn 1977, 322). But he still claims that even if we were to universally accept these five criteria, scientists "fully committed to the same list of criteria for choice may nevertheless reach different conclusions" (Kuhn 1977, 322).

Aiming to address the Kuhnian worry and defend scientific realism, Sam Schindler's recent book *Theory Virtues in Science: Uncovering Reality through Theory*, also advances arguments to show how theory virtues can justify our belief in the truth of a theory. Schindler calls the argument from the variation and instability of virtues, developed by Duhem and Kuhn "the conflict thesis", and takes it to be an empirical fact that this conflict occurs in science. He then proceeds to advance his "no virtues coincidence argument", taken to be the ultimate defence for scientific realism. Schindler's starting point is to acknowledge that the most widely accepted defence of scientific realism—the no miracles argument—suffers from the base rate fallacy. As Magnus and Callender (2004) outline, for the no miracles argument to be sound, the prior probability of the truth of the theory in consideration must be sufficiently high, a premise that the empiricist or instrumentalist would not grant. Taking this as a detrimental objection against realism, Schindler turns to theory virtues for a defence of scientific realism. Starting from the premise that theory virtues are diverse and interpreted in a plurality of ways, Schindler claims that the very fact that sometimes there is convergence between scientists on the merits of a theory gives an argument for scientific realism. He argues "it would be an unlikely coincidence for scientists' judgement concerning the truth of a theory on the basis of the theory's virtues to converge if the theory wasn't in fact true" (Schindler 2018, 52). According to Schindler the no virtue coincidence argument offers the ultimate defence of realism, explaining that

> if the conflict thesis is true and there are only sometimes theories that exhibit several or even all of the standard virtues, then it would be a strange coincidence if a theory had all the five virtues, was embraced by all scientists, and was not true. It is easy to see: just like the standard NMA for realism, this is a no-coincidence argument. (Schindler 2018, 55)

Similar convergence arguments have been employed in other contexts. For instance, Hacking (1983) argues that when multiple different methods of measurement of instruments produce the same results, we have grounds to infer they are caused by the same entity. For instance, Jean Perrin famously derived 13 ways of calculating the number of molecules in a mole of substance, which he publicized between 1908 and 1913 in his books on *Brownian Movement and Molecular Reality* and *Atom*. Many have interpreted this argument as a "no coincidence argument", claiming that it would be unlikely coincidence for the determinations to be found in such agreement if the atomic hypothesis was not indeed correct and there are not such particles as molecules and atoms.[3] This result gives reason for accepting the atomic hypothesis and concluding a long lasting debate about the reality of the atom. In a similar vein, Schindler argues that it is highly unlikely that scientists, who have very different rankings and definitions of theory virtues, would accept a theory unless this theory is true. Schindler suggests that the defence of realism advanced is superior to the standard no miracles argument for scientific realism because it is not subject to the base rate fallacy. The problem is avoided, he argues, since the importance of the assigned prior probability of the theory is suppressed, since the posterior probability approaches one as the number of scientists accepting a theory as true increases.

Schindler's argument has given new life to the realism debate and the role of theory virtues in theory acceptance, but how plausible is it? One difficulty for the no virtue coincidence argument is that it relies on the premise that the scientific community's criteria of acceptance (the theory virtues) would be rather diverse, which even a Kuhnian would challenge, since the virtues of a good theory do not tend to deviate much. As we saw earlier, Kuhn himself settled the five theory virtues of simplicity, scope, fruitfulness, accuracy and consistency. James McAllister, who sees theory change as a revolutionary transition of aesthetic canons, expecting diversity between communities on what constitutes a beautiful theory and allows beauty canons to be dynamic and evolving, argues that as a matter of fact there is substantial overlap between communities as to what aesthetic properties of theories are valued. Furthermore, Montano (2016) and Ivanova (2020, 2023) argue that some virtues are central in science practice and do not seem to change in theory transition. Montano calls virtues such as simplicity and scope "historical constants", arguing that they are consistently sought after and Ivanova (2020) explains the permanence of such virtues as grounded in facts about cognition and ways in which scientists as agents process information. To sum up, the no coincidence argument presupposes a diversity of virtues, which seems hard to motivate by reflecting on the practice of science.

The above arguments aim to establish that there is an underlying relationship between theoretical virtues and likelihood of truth. We have discussed some difficulties for these accounts and found them to be unconvincing. In the next section I want to return to van Fraassen's insight from the *Scientific Image* regarding the status of theoretical virtues seen as pragmatic devices guiding our choices, reflecting our interests and aims. I want to offer further support to this claim by looking

---

3 The no coincidence interpretation of Perrin's experiments is not uncontroversial. See van Fraassen (2014) for an alternative interpretation.

into how we process information and how theoretical virtues can impact on our inferential practices.

4 Aims in Science and Place for Theory Virtues

In the *Scientific Image* van Fraassen stresses the importance of agents, their aims and how their interests affect decision-making. In this section I want to focus on some empirical evidence from cognitive science that supports van Fraassen's argument, claiming that the theory virtues we so much value in our theories are reflections on our cognitive capacities and limitations as agents, not a reflection of how the world is. Science today, whether we look into physics, biology, or chemistry, gives us ground to doubt that the phenomena are simple, unified, nicely reduce to a fundamental level, and that a theory of everything is even possible. But scientists are often driven by such ideals, aspire for simple explanations, elegant descriptions of reality, and in some areas of science there is still underlying hope for a grand unifying theory. Such considerations are important and heuristically useful, but they seem to reflect our own intellectual capacities. Here I discuss two kinds of studies to support this conclusion, work on how theory virtues constrain inferential reasoning and studies on processing fluency.

Recent work in cognitive psychology has focused on studying agent's preferential inferences and supports the idea that agents strive to develop and consistently support simpler explanations and explanations of broader scope. Tania Lombrozo discusses numerous studies that illustrate how our reasoning is constrained by virtues such a simplicity and breadth, with test subjects consistently preferring hypotheses of broader scope and simple form. In the noted studies subjects, both adults and children, employed such theory virtues as simplicity and scope in their choice of hypotheses. Interestingly these virtues affected the kind of patterns the subjects were able to discover as well as what hypotheses they were able to come up with. This observation is very much in line with an insight from Henri Poincaré, who argues that the regulative role simplicity and unity play in our theorizing about the world affect our selection of facts to study and the generation of hypotheses. Lombrozo makes the interesting observation that even when the subjects in the studies were given base rates for the hypotheses they were choosing between, so they were able to not only appreciate the relevant simplicity and scope of an explanation but also its likelihood, it was not until the complex explanations were significantly more likely that the subjects changed their mind:

> In one set of studies, participants selected between explanations for an individual's symptoms that appealed to either a single common cause (simple) or to two independent causes (com-

> plex), and they were additionally provided with information about the base rate of each disease. Under these conditions, explanation choices were sensitive to both simplicity and probability. When two candidate explanations were equally likely, a majority of participants selected the simpler explanation. It was not until the complex explanation was 10-fold more likely than the simpler alternative that a majority of participants selected it as the most satisfying explanation for the symptoms. It thus appears that, when the probability of an explanation is uncertain, the relative simplicity of an explanation has a significant effect on its perceived quality. (Lombrozo 2016, 751)

The tendencies to prefer simple and unified hypotheses highlighted in these studies can serve to explain why scientists can often seem resistant to revise their attitudes towards theory virtues and continue to prefer theories that satisfy their ideals of beauty even when the data would not support them.

The connection between our cognitive make-up and our preference for a certain type of hypotheses has also been explored by reflecting on processing fluency. Numerous studies in cognitive science have highlighted a connection between processing fluency and the trust agents place in a hypothesis. The studies reveal that a person's trust in a hypothesis is influenced by how the information is delivered, with a diversity of factors being potentially responsible for processing fluency, with hypotheses being perceived as more trustworthy if they were easy to visualize, delivered by a trustworthy person, or someone who is familiar, speaks clearly or in a familiar accent (Schwarz 2018). Processing can be fluent due to the features of the subject matter—a simple theorem, an elegant mathematical proof, a clearly written essay are easier to process, but even the font and colour in which the information is presented, as well as whether the idea is accompanied by a visual representation or familiar example affects fluency. In addition, processing fluency is related to the exposure effect. As studies have found, the more one is exposed to a particular idea, artwork or theory, the easier it is to process it and find it beautiful. Studies on exposure effect have shown that exposure is crucial to aesthetic appreciation, and that even if subjects are initially indifferent to a particular artwork or music, repeated exposure increases their aesthetic appreciation of it (Zajonc 1968; Cutting 2003). James McAllister (1996) has argued that this is the mechanism responsible for scientist's aesthetic judgements—some theories are deemed beautiful after prolonged exposure to working with these theories.

The relationship between processing fluency and judgements of truth is important to stress. The above-mentioned studies support the idea that people tend to trust a hypothesis when this hypothesis is easy to understand and this phenomenon occurs also in conjunction with the exposure effect. Some studies have observed how people's trust in a rumour increases with the number of times they have heard it. This phenomenon can explain the claim often made by scientists that beautiful theories are true. Such theories exemplify features, such as simplicity, harmony, symmetry, visualization, that contribute to their processing, making the subject more likely to judge the theory to be true. But what we can conclude from these studies is not that such theories are in fact more likely to be true, but simply that such theories are easy for us to understand and reflect our cognitive make-up.[4]

Departing from the traditional aims of science, found in the acquisition of truth or empirical adequacy, some philosophers have recently turned to the notion of understanding construed as an independent epistemic goal of science. In the work of Henk de Regt (2004) and Catherine Elgin (2009, 2017), for example, understanding is construed contextually and independently of the notion of truth, in the sense that truth is not a necessary condition for understanding. This move is motivated by a desire to accommodate for how science operates, specifically how we can gain understanding of worldly phenomena and develop successful theories that enrich our understanding by often making knowingly false assumptions. Scientific models regularly distort and misrepresent their target systems, but they allow us to preform successful epistemic activities and gain understanding of phenomena. This focus on action and practice rather than propositional knowledge does not take understanding to be a kind of knowledge but rather as a skill or ability to grasp how the facts fit together. This literature has opened the door to showing how theory virtues relate to achieving understanding. The idea that theory virtues guide understanding can be found in recent literature in philosophy of science, advocated for instance in Breitenbach (2013), de Regt (2009), Ivanova (2020), Kosso (2002), Elgin (2020), and even in pre-positivistic philosophers, such as Henri Poincaré (Ivanova 2017). Importantly, some of these authors argue against the relationship between truth and theory virtues and rather propose that theory virtues be seen as vehicles or regulative ideals for the achievement of understanding.

Perhaps focusing on understanding rather than truth holds promise for those who want to assign some epistemic significance to theory virtues while staying away from the notion of truth. Appealing to contextual accounts of understanding according to which understanding is expressed as a skill or ability to perform certain epistemic activities allows us to focus on the role played by theory virtues. Our preference for elegance or unity of a theory reflects not certain features of the world, but rather our choice to construct the theory in such a way because a theory that fits these ideals better suits our aims. Taking these virtues as conditions of our cognitive make-up explains why certain theory virtues persist even when our best theories fail to exhibit them.

4 For a discussion of processing fluency and agent understanding, see Arcangeli and Dokic (2020) and Todd (2020).

5 Conclusion

The debate on the aims of science has been shaped by Bas van Fraassen's development of constructive empiricism, offering a strong alternative to realism. According to constructive empiricism, the aim of science is to provide empirically adequate theories and commitment to unobservable entities is unnecessary; rather theory acceptance equates with belief in the empirical adequacy of a theory. When it comes to theory acceptance on grounds other than empirical adequacy, van Fraassen's contribution is to ask what evidence we have in trusting that the non-empirical factors such as theory virtues can be evidential and reveal the likelihood of a theory to be true. We not only lack good evidence that theory virtues are truth-tracking but we also have good evidence that theory virtues are importantly subjective, and often our own intellectual capacities and limitations can make us select and favour theories that satisfy these requirements. In this chapter I have elaborated on this insight and supported the idea that the virtues often employed in scientists' decision-making are grounded in their aims, capacities and cognitive limitations and cannot be taken to reflect the likelihood of a theory to be successful. It is certainly a fact that scientists use theory virtues in the conception and evaluation of theories and they can credit theories on non-empirical grounds, but these practices seem to reflect not facts about how the world is but rather the preferences, aims and pragmatics of the agents involved in selecting facts, inferences and frameworks within which to develop theories.

References

Arcangeli, M. and Dokic, J. 2020. A Plea for the Sublime in Science. In: *The Aesthetics of Science: Beauty, Imagination and Understanding*, eds. S. French and M. Ivanova, 104–124. New York: Routledge.

Breitenbach, A. 2013. Aesthetics in Science: A Kantian Proposal. *Proceedings of the Aristotelian Society* 113: 83–100.

Bueno, O. et al. 2017. *Thinking about Science, Reflecting on Art: Bringing Aesthetics and Philosophy of Science Together.* New York: Routledge.

Chandrasekhar, S. 1987. *Truth and Beauty: Aesthetics and Motivation in Science.* Chicago: The University of Chicago Press.

Cutting, J. 2003. Gustave Caillebotte, French Impressionism, and Mere Exposure. *Psychonomic Bulletin & Review* 10: 319–343.

De Regt, H. W. 2004. Discussion Note: Making sense of understanding. *Philosophy of Science* 71(1): 98–109.

De Regt, H. W. 2009. The Epistemic Value of Understanding. *Philosophy of Science* 76(5): 585–597.

Dirac, P. A. M. 1980. The Excellence of Einstein's Theory of Gravitation. In: *Einstein: The First Hundred Years*, eds. M. Goldsmith, A. Mackay and J. Woudhuysen, 41–46. Oxford: Pergamon Press.
Duhem, P. 1954 [1906]. *The Aim and Structure of Physical Theory.* Princeton: Princeton University Press.
Elgin, C. Z. 2020. Epistemic Gatekeepers: The Role of Aesthetic Factors in Science. In: *The Aesthetics of Science: Beauty, Imagination and Understanding*, eds. S. French and M. Ivanoa, 21–35. New York: Routledge.
Elgin, C. Z. 2017. *True Enough.* Cambridge, MA: MIT Press.
Elgin, C. Z. 2009. Is Understanding Factive? In: *Epistemic Value*, eds. D. Prichard, A. Millar and A. Haddock, 322–330. Oxford: Oxford University Press.
French, S. and Ivanova, M. 2020. *The Aesthetics of Science: Beauty, Imagination and Understanding.* New York: Routledge.
Heisenberg, W. 1971. *Physics and Beyond: Encounters and Conversations.* Trans. by Arnold J. Pomerans. London: George Allen and Unwin.
Hossenfelder, S. 2018. *Lost in Math: How Beauty Lead Physicists Astray. Basic Books.* Cham: Springer International Publishing.
Ivanova, M. 2010. Pierre Duhem's Good Sense as a Guide to Theory Choice. *Studies in the History and Philosophy of Science* 41: 58–64.
Ivanova, M. 2017a. Aesthetic Values in Science. *Philosophy Compass* 12. DOI: 10.1111/phc3.12433.
Ivanova, M. 2017b. Poincaré's Aesthetics of Science. *Synthese* 194(7): 2581–2594.
Ivanova, M. 2020. Beauty, Truth and Understanding. In: *The Aesthetics of Science: Beauty, Imagination and Understanding*, eds. S. French and M. Ivanova, 86–103. New York: Routledge.
Ivanova, M. 2021. *Duhem and Holism.* Cambridge: Cambridge University Press.
Ivanova, M. 2023. What Is a Beautiful Experiment? *Erkenntnis.* **88**, 3419–3437.
Ivanova, M and Murphy, A. 2023. *The Aesthetics of Scientific Experiments*. New York: Routledge.
Kaku, M. and Thompson, J. T. 1997. *Beyond Einstein: The Cosmic Quest for the Theory of the Universe.* Oxford: Oxford University Press.
Kosso, P. 2002. The Omniscienter: Beauty and Scientific Understanding. *International Studies in the Philosophy of Science* 16: 39–48.
Kuhn, T. S. 1977. Objectivity, Value Judgment, and Theory Choice. In: T. Kuhn: *The Essential Tension*, 320–353. Chicago: The University of Chicago Press.
Kuhn, T. S. 1996 [1962]. *The Structure of Scientific Revolutions.* 3rd ed. Chicago: University of Chicago Press.
Mach, E. 1984. *The Analysis of Sensations and the Relation of the Physical to the Psychical.* Trans. by C. M. Williams. La Salle: Open Court.
Magnus, P. and Callender, C. 2004. Realist Ennui and the Base Rate Fallacy. *Philosophy of Science* 71: 320–338.
McAllister, J. 1996. *Beauty and Revolution in Science.* Ithaca, NY: Cornell University Press.
Montano, U. 2014. *Explaining Beauty in Mathematics: An Aesthetic Theory of Mathematics.* Synthese Library 370, Cham: Springer.
Newton, I. 1999 [1687], *The Principia: Mathematical Principles of Natural Philosophy.* Trans. by I. B. Cohen and A. Whitman. Berkeley: University of California Press.
Lombrozo T. 2016. Explanatory Preferences Shape Learning and Inference. *Trends Cogn Sci.* 20(10):748–759. DOI: 10.1016/j.tics.2016.08.001.
Poincaré, H. 2001 [1902, 1905, 1908]. Science and Hypothesis. In: *The Value of Science: Essential Writings of Henri Poincaré*, ed. S. Gould, 1–178. New York: Modern Library.
Psillos, S. 1999. *Scientific Realism: How Science Tracks Truth.* London: Routledge.

Salmon, W. 1990. The Appraisal of Theories: Kuhn Meets Bayes. *PSA: Proceedings of the Biennial Meeting of the Philosophy of Science Association* 1990, Vol. 2: 325–332.

Schindler, S. 2018. *Theoretical Virtues in Science: Uncovering Reality Through Theory.* Cambridge: Cambridge University Press.

Schwarz, N. 2018. Of Fluency, Beauty, and Truth: Inferences from Metacognitive Experiences. In: *Metacognitive Diversity: An Interdisciplinary Approach*, eds. J. Proust and M. Fortier, 25–46. Oxford: Oxford University Press.

Stump, D. 2007. Pierre Duhem's Virtue Epistemology. *Studies in History and Philosophy of Science* 38: 149–159.

Swinburne, R. 1997. Simplicity as Evidence of Truth. Milwaukee: Marquette University Press.

Todd, C. 2020. Imagination, Aesthetic Feelings, and Scientific Reasoning. In: *The Aesthetics of Science: Beauty, Imagination and Understanding*, eds. S. French and M. Ivanova, 63–85. New York: Routledge.

van Fraassen, B. 1980. *The Scientific Image.* Oxford: Clarendon Press.

van Fraassen, B. 2009. The Perils of Perrin, in the Hands of Philosophers. *Philosophical Studies* 143(1), 5–24.

Zajonc, R. B. 1968. Attitudinal Effects of Mere Exposure. *Journal of Personality and Social Psychology: Monograph Supplement* 9: 1–27.

Zeki, S., Romaya J. P., Benincasa, D. M. T. and Atiyah M. F. 2014. The Experience of Mathematical beauty and Its Neural Correlates. *Frontiers in Human Neuroscience* 8: 1–12.

Michael Friedman

Realism and Instrumentalism: The Best of Both

Abstract: I follow Howard Stein's project of reconciliation between two attitudes towards scientific theories that are standardly taken to be diametrically opposed to one another—realism and instrumentalism. Stein argued that realism and instrumentalism are fruitfully combined with one another in the work of the greatest scientists, and, in particular, in the work of Newton, Maxwell, and Einstein. These works entail realism in so far as their goal is adequately and correctly "to represent the phenomena of nature". However, they entail instrumentalism also, in so far as Newtonian methodology explicitly includes the possibility (and even probability) that any theory with justifiable claims to truth (e.g., Newtonian universal gravitation) can eventually lead to a quite distinct theory with even better such claims to truth (Einsteinian general relativity).

1 Introduction

Realism Today occupied the 8th Colloquium of the Institut de la Méthode at a short distance from Bern in Neuchâtel. These colloquia were funded and organized by Henri Lauener (13 July 1933 to 28 October 2002), a Swiss philosopher interested in both Continental and Analytic approaches. He edited the journal *Dialectica* and established the Lauener Foundation for Analytical Philosophy, which supported (and continued to support) a biennial prize for honouring a leading philosopher of the time.

The 8th Colloquium took place on 26–29 May 1988, appearing in *Dialectica* in 1989. The theme of *Realism Today* was prominently represented by John Worrall's "Structural Realism: The Best of Both Worlds?" But the other contender was Howard Stein's "Yes, But..., Some Skeptical Remarks on Realism and Anti-Realism". My own view is more like Stein's genuinely sceptical approach, while Worrall's "Structural Realism" is often taken to be a one-sidedly realist view—with the question mark at the end indicating some evidence of confusion. I will take my own stand with Stein, but there are more than enough "realist" and "anti-realist" views needing serious clarification.

Bas van Fraassen won the biennial prize 20 years later in 2018. I confess that I did not do him justice when I was younger, and it took me many years to appreciate the fact that Bas has formulated a clear philosophical stance for himself—one

https://doi.org/10.1515/9783111019802-011

that is definitely closer to "instrumentalism" and "anti-realism". I am therefore glad that I have been invited to the present Colloquium so that I can now make amends. My point, in the end, amounts to Stein's "Skeptical Remarks", whereby we accept "realism" but embrace what Stein calls "instrumentalism" at the same time.[1]

I have suggested, with hindsight, that the paper most influencing the course of the scientific realism debate was Worrall's paper, whose full title was "Structural Realism: The Best of Both Worlds?" The two worlds in question were realism and anti-realism, and Worrall attempted to do justice to both (if only somewhat tentatively) by reviving what he took to be an earlier version of structural realism articulated by Henri Poincaré. Worrall's idea, in brief, was to find sufficient continuity in scientific change to support the legitimate claims of scientific realism while simultaneously acknowledging that the referents of central theoretical terms may radically change in the process or even disappear altogether. His main example involved the mechanical models of the luminiferous and electromagnetic aether that were explored throughout the nineteenth century—which, it is supposed, were later found to have no physical reality after all, especially in the context of Einsteinian relativity theory. Worrall's contention, in a nutshell, was that, although these aethers themselves conceived as material substances, were not preserved over time, the mathematical structures involved in their description—wave forms continuously propagating over space and time, electromagnetic lines of force (vector fields) in the theories of Faraday and Maxwell—were preserved nonetheless. It is here that we find the invariant *mathematical structures* that remain after the disappearance of their originally-supposed *material content.*

The full title of Stein's paper is "Yes, But…, Some Skeptical Remarks on Realism and Anti-Realism". Here, without referring to Worrall's paper directly (note 59 above), Stein embarked on a seemingly analogous project of reconciliation between two attitudes towards scientific theories that are standardly taken to be diametrically opposed to one another—this time realism and instrumentalism. Stein argued, on the contrary, that realism and instrumentalism are fruitfully combined with one another in the work of the greatest scientists, and, in particular, in the work of Newton, Maxwell, and Einstein: realism in so far as their goal is adequately and correctly "to represent the phenomena of nature", but instrumentalism also, in so far as Newtonian methodology explicitly includes the possibility (and even

1 It is interesting to note that the printed version of Worrall (1989) explicitly thanks Stein for his helpful comments. It is also interesting to note that Stein states his conclusion as follows (1989, 65): "Realism—Yes, but instrumentalism—Yes also, No only to anti-realism." Stein's substitution of "instrumentalism" for "anti-realism" can be taken to signal an implicit disagreement with the basic approach of Worrall's paper.

probability) that any theory with justifiable claims to truth (e.g., Newtonian universal gravitation) can eventually lead to a quite distinct theory with even better such claims to truth (Einsteinian general relativity).

Now this talk of representing the phenomena, and of finding abstract theoretical structures in the phenomena, may appear unsatisfactory from a realist point of view. For one of the key moments in the development of modern science—Galileo's refusal to take up Cardinal Bellarmine's invitation to confine Copernican theory to the goal of "saving the phenomena"—appears to have involved precisely such a realist commitment going well beyond this goal. And a similar dialectic, of course, appears to be present in later collisions between realism and instrumentalism, such as Pierre Duhem's rejection of contemporaneous molecular-atomic theories (again on behalf of "saving the phenomena"). I confess that I was not quite satisfied with Stein's position, along precisely these lines for a considerable period of time, especially in my earlier militantly realist phase. Moreover, I continued to have trouble understanding Stein's position even as I became ever more Kantian, despite the fact that Stein explicitly appeals to Kant's notion of phenomena in his criticism of what he takes to be *excessive* realism.[2] I think that I have now found the source of my misunderstanding, and it is interestingly connected with a second misunderstanding of an apparent criticism of Newton articulated by Kant.

2 Kant and Newton on Force and Quantity of Matter

Kant's apparent criticism of Newton appears in the second remark to Proposition 7 in the second chapter of the *Metaphysical Foundations of Natural Science* (1786),

2 In defending his conception of adequately and correctly representing the phenomena against a form of realism demanding something more (reference to something "behind" the phenomena), Stein begins (1989, 50): "First, there is the classical philosophical problem of the epistemology of transcendent metaphysics—the problem that Kant (we may say) posed to the ghost of Leibniz, and that led Kant himself to the view that all theoretical knowledge is *of* the phenomenal world: the world of experience. The nub of the matter can be put very simply: it is, How can you *know* that things are as you say they are? If the claimed 'reference' of the theory is something beyond its correctness and adequacy in representing the phenomena—if, that is, for a given theory, which (we may suppose) does represent phenomena correctly and adequately, there are still two possibilities: (a) that it is (moreover) *true*, and (b) that it is (nonetheless) *false*—then how in the world could we ever tell what the actual case is?" This Steinean passage, for me, reveals a deep connection between Stein's approach and that of Kant—thereby justifying my pairing of the two throughout this paper.

where Kant appears sharply to differ with Newton concerning gravity as an immediate action at a distance (AA 4, 515):

> [I]t is clear that the offense taken by his contemporaries, and perhaps even by Newton himself, at the concept of an original attraction set him at variance with himself. For he could by no means say that the attractive forces of two planets, those of Jupiter and Saturn for example, manifested at equal distances of their satellites (whose mass is unknown), are proportional to the quantity of matter of these heavenly bodies, if he did not assume that they attracted other matter merely as matter, and thus according to a universal property of matter.[3]

What is at issue is the method Newton develops in Book 3 of the *Principia* for measuring the quantities of matter of primary bodies in the solar system by the accelerations produced in their satellites. Kant's point appears to be that Newton must assume both the universality and the immediacy of the gravitational interaction between Jupiter and Saturn independently of any intervening matter surrounding these bodies. Kant thus appears to insist on the physical reality of immediate action at a distance for gravitational force, contrary to Newton's own more agnostic approach.

Since the Newtonian method for measuring the quantities of matter of primary bodies in the solar system is a paradigmatic example of what G. E. Smith calls theory-mediated measurement, it is well worth our while to examine this method more closely.[4] The first point to emphasize is that these measurements enabled Newton empirically to demonstrate that the centre of gravity of the solar system lies always very close to the centre of the Sun (Book 3, Proposition 12). Newton was considering, in particular, the choice between the Copernican and Tychonic systems, having already eliminated the Ptolemaic system via telescopic observations (following Galileo) of the phases of Venus and Mercury. Newton could then measure the quantity of matter of the Sun by the orbital acceleration of Venus (the same on both Copernican and Tychonic systems), the quantities of matter of Jupiter and Saturn by the orbital accelerations of their satellites (likewise), and that of the earth by the orbital acceleration of our moon (the same on all three systems). It thereby turned out that the quantity of matter of the earth is much, much smaller than that of the Sun, so that only the Copernican system (in its Keplerian formulation) remained. And it is precisely this argument that has settled the empirical truth of the Copernican system (to a very high degree of approximation) ever since.

---

3 I cite this work by volume and page numbers from the Academy edition. The page numbers in vol. 4 appear in the margins of my contribution to the Cambridge Edition (Kant 2004 [1786]). All translations from Kant's German are my own.

4 Smith (2002) is a valuable introduction to his conception of Newtonian methodology.

How did Newton arrive at his method for measuring the quantities of matter of the primary bodies in the solar system? Two propositions established earlier in Book 3 are of particular importance. Proposition 6 generalizes the (Galilean) equality of the acceleration of fall for all bodies near the surface of the earth to the gravitational fields of any "planet" (i.e., primary body), now considered as extending to arbitrary distances (Newton 1999, 806):

> All bodies gravitate toward each of the planets, and at any given distance from the center of any one planet the weight of any body whatever toward that planet is proportional to the quantity of matter which the body contains.[5]

Here we are dealing with the singular property of gravitational force that it produces the same acceleration in any body whatsoever at equal distances from the source of that force—entirely independently, in particular, of the accelerating body's mass. This property does not hold for other forces such as magnetism, and it amounts to the claim that the specifically gravitational force on an attracted body (i.e., its "weight") is always proportional (at a given distance) to this body's mass, and it therefore essentially involves what we now call the equality of inertial and *passive* gravitational mass.[6]

This property, the (generalized) Galilean property of the gravitational field, is a crucial part of Newton's theory-mediated measurements of the quantities of matter of the primary bodies in the solar system. As becomes clear in Newton's following Proposition 7, however, the Galilean property alone is not sufficient. According to Proposition 7 (1999, 810): "Gravity exists in all bodies universally and is proportional to the quantity of matter in each." The proof depends on Proposition 69 of Book 1, which appeals to the Galilean property (the equality of inertial and passive gravitational mass), the universality of the attractive force in question (in this case gravitation), and, crucially, the equality of action and reaction expressed in the Third Law of Motion. Consider the system of bodies consisting of Jupiter, Saturn, and their satellites. By the Galilean property there are what Stein (1967) has helpfully called "acceleration fields" around the two planets, whose actions are independent of the mass of any body "falling" in such a field.

---

5 All citations to the *Principia* are from Newton (1999).

6 Inertial mass, a concept introduced by Newton for the first time in Definition 3 of the *Principia*, is a measure of the resistance of a body to motion (acceleration) produced by *any* force, encapsulated in our modern equation $F = ma$. Gravitational mass, by contrast, represents the source of the gravitational field, so that, in our modern formulation, $F_{grav} = Gm_A m_B/r^2$. If A is the attracting body and B the attracted body, we then have $a_B = Gm_A/r^2$. The *passive* gravitational mass of the attracted body, taken to be equal to its inertial mass, drops out of the equation.

By universality, these fields extend (in the inverse-square proportion to distance) arbitrarily far into space and, in particular, between the two planets themselves. Applying the equality of action and reaction to this interaction implies that the acceleration produced by Jupiter on Saturn, for example, is proportional (at the given distance) to the *inertial* mass of Jupiter (and conversely for Saturn). By measuring the acceleration of one of Jupiter's satellites in the same acceleration field, we are therefore measuring the inertial mass of Jupiter as well (and similarly for Saturn and its satellites). In other words, the equality of inertial and passive gravitational mass, in the context of the universality of gravitational interaction and the Third Law of Motion, implies the equality of inertial and *active* gravitational mass as well.[7]

In order better to appreciate the subtlety and complexity of the Newtonian measurement of quantity of matter we need also to consider the traditional concept of *weight* from (Archimedean) statics in relation to both the distinctively Newtonian concept of inertial mass and the concept of quantity of matter that was already in use before Newton. Indeed, what Newton makes of the concept of weight is the key to his unification of the other two concepts. (Terrestrial) weight, for Newton, is the force on a body exerted by the earth's gravitational field, while inertial mass, as we have seen, represents the resistance of a body to *any* force. Given the Galilean property that all bodies fall with (approximately) the same acceleration near the surface of the earth (the constant *g* in Galileo's law of fall), it follows that two falling bodies are in equilibrium (in a balance, for example) just in case their (Newtonian) inertial masses are equal. Inertial mass, in this context, represents what the weights of bodies near different sources of attraction (the earth and the Sun, for example) have in common. In one use of the concept of quantity of matter, therefore, inertial mass represents the invariant or intrinsic

7 Let the acceleration field on Saturn's moons be given by $a_1 = k_S/r_1^2$ and the acceleration field on Jupiter's moons by $a_2 = k_J/r_2^2$, where k_S and k_J are constants characterizing the absolute strengths of the fields independently of distance. Since the two acceleration fields extend far beyond the regions of their satellites, we also have accelerations $a_J = k_S/r^2$ of Jupiter and $a_S = -k_J r^2$ of Saturn, where r is the distance between them. But, according to the Third Law, $m_J a_J = -m_S a_S$. Therefore, $m_S/m_J = -a_J/a_S = k_S/k_J$: the absolute strengths of the two fields—their *active* gravitational masses —are proportional to their inertial masses. The need for this second step (Proposition 7) is obscured by the modern formulation of the previous fn., for it already assumes that there is a universal constant G characteristic of all gravitational interactions such that $k_A = Gm_A$, where m_A is the inertial mass of the attracting body. Given the conceptual distinction between inertial and gravitational mass, however, this is precisely what Newton needs to establish—first via the acceleration-field property for passive gravitational mass and then by the above argument for active gravitational mass.

quantity of matter that a body has independently of its interactions with other bodies.

In the seventeenth century, however, there was also another use of the concept of an intrinsic or invariant quantity of matter defined independently of weight: the amount of "stuff" in a given volume occupied by matter, which could be made more precise in terms of the ratio between the volume of absolutely dense (incompressible) matter in the volume (without empty interstices) and the total volume. This purely geometrical definition of quantity of matter was popular among adherents of the mechanical natural philosophy, and it, too, had a well-defined mathematical meaning independently of a body's interactions and motions relative to other bodies (including its merely potential motions as in the concept of inertial mass). The problem was that this purely geometrical concept had no clear physical or empirical meaning at the time: how in the world could the ratio between absolutely dense matter within a given volume and the total volume be measured, even in principle? In that situation, therefore, we had a mathematically well-defined quantity that was not, at the same time, empirically well-defined (empirically measurable). This underscores and makes vivid the importance of theory-mediated measurement.

If we adopt Newtonian inertial mass as the invariant concept of quantity of matter corresponding to the measurement of weight, however, the remaining problem is to explain how these measurements can then be projected into the heavens so as to apply to bodies beyond the significant reach of the earth's gravitational field. Such bodies can certainly not be quantitatively compared by placing them in a balance, for example. Here we find one of the most interesting and subtle steps in Newton's argument, which occupies Propositions 3–4 of Book 3, including the Scholium to Proposition 4. This argument, featuring what we now call the moon test, follows Propositions 1 and 2 of Book 3, which have established the inverse-square proportion of acceleration for the moons of Jupiter and Saturn relative to their primary bodies, and the same proportion for the primary planets (excepting the earth) relative to the Sun. Since there are a number of different orbiting bodies in each case, the inverse-square proportion, by Kepler's third or harmonic law, applies from orbit to orbit (not just within a single orbit). And since the harmonic law depends only on distance and orbital period (independently, crucially, of the masses of the orbiting bodies) the acceleration-field property therefore follows.[8]

8 Assuming circular orbits for simplicity, and using the circumstance that here $v = 2\pi rt$, with r the radius and t the period, we have $a = v^2/r$ in accordance with the formula for centripetal (or centrifugal) acceleration. In these terms, therefore, Kepler's harmonic law states that if r, R and t,

In the case of the (earth's) moon considered in Proposition 3, however, there is only a single orbiting body. So we cannot appeal directly to Kepler's harmonic law in order to derive the acceleration-field property. Newton, in Proposition 3, does not even mention the harmonic law in deriving the inverse-square proportion, but appeals, instead, to "the very slow motion of the moon's apogee" (1999, 802).[9] This derivation of the inverse-square proportion, however, holds only for the (single) lunar orbit and thus does not by itself imply the acceleration-field property. What we now call the moon test, which is suggested in the Corollary to Proposition 3 and then carried out in Proposition 4, therefore proceeded as follows. Beginning from the inverse-square proportion for the (single) lunar orbit, Newton compared the (actually measured) acceleration of the orbiting moon towards the centre of the earth with the acceleration that the moon would have if it were deprived of its orbital motion and to descend to the surface of the earth via a continual increase of this acceleration in accordance with the inverse-square proportion. It turns out, on this supposition, that the resulting acceleration at the surface of the earth would equal the Galilean acceleration g of terrestrial gravity (as determined, for example, by the motions of pendula). This gave Newton licence to conclude, in Proposition 4, that the force by which the moon is maintained in its orbit is the very same force as terrestrial gravity, and also, in the Scholium to this proposition, that the relationship between falling bodies near the surface of the earth and the moon is precisely analogous to that between the different (multiple) satellites of the (known) primary bodies in the solar system (the satellites of Jupiter, Saturn, and the Sun). We now (and only now) have the acceleration-field property of gravitation in *all* of these cases.[10]

T are, respectively, the distances and periods of two concentric orbits, then $r^3/R^3 = t^2/T^2$. It follows that the two accelerations stand in the ratio $1/r^2:1/R^2$ and thus that the variation in acceleration depends only on distance in accordance with the inverse-square proportion.

9 Newton had already indicated this way of deriving the inverse-square proportion, by way of anticipation, in Proposition 2, with reference to Corollary 1 of Proposition 45 of Book 1. In Proposition 2, however, Newton says that the inverse-square proportion can be "proved with the greatest exactness from the fact that the aphelia are at rest" (1999, 802), while in Proposition 3 he appeals to a corresponding "very slow" motion of the lunar orbit—indicating, in this case, that the inverse-square proportion holds only to a very high degree of approximation.

10 The main argument of Proposition 4 concludes as follows (1999, 804): "And therefore that force by which the moon is kept in its orbit, in descending from the moon's orbit to the surface of the earth, comes out equal to the force of gravity here on earth, and so (by rules 1 and 2) is that very force which we generally call gravity." The Scholium begins (805): "If several moons were to revolve around the earth, as happens in the system of Saturn or of Jupiter, their periodic times (by the argument of induction) would observe the law which Kepler discovered for the planets [i.e., the harmonic law], and therefore their centripetal forces would be inversely as the squares of the distances from the center of the earth (by prop. 1 of this book 3)." Note that the original Gal-

What has been called the "Newtonian Style", first expressed by I. B. Cohen in *The Newtonian Revolution* (1980), consists in a series of *approximative corrections*, moving from one-body systems, to two-body systems, and finally to three- and many-body systems. G. E. Smith adds the concept of *theory-mediated measurements* to the concept of approximative corrections.[11] What makes the relevant notion of "force" (in approximative corrections and theory-mediated measurements) *abstract* becomes clear in Propositions 1–2 of Book 1, where mathematical properties of forces can be derived from those of the corresponding motions. And, in particular, Kant accepts Newton's abstract treatment of impressed force as a measurable mathematical magnitude independently of any hypotheses concerning underlying mechanisms. We must then show that gravitational force, in particular, has the mathematical properties required in order to facilitate theory-mediated measurements of both this force itself and other centrally important related quantities. Prominent among the latter is quantity of matter, the measurement of which, in the theory of universal gravitation, involves the direct application of action = reaction (conservation of momentum) between arbitrarily distant bodies.

This kind of abstract "structural" conception is not a version of contemporary "structural realism". Rather, it is important to Howard Stein that his conception involves *both* "realism" and "instrumentalism" (see notes 59 and 60). It concerns, in my view, the *methodological* use of theory-mediated measurements in ongoing inferences that are more secure than either inductive (curve fitting) or hypothetico-deductive arguments (see I. B. Cohen and G. E. Smith above). The greater security of such inferences is a centrally important part of Kant's conception of causal necessity. And these inferences are subject to later refinements along new and unanticipated dimensions that may become clear only post hoc. Kant is less explicit about —and in the end less open to—such Newtonian refinements. Nevertheless, there is at least one explicit reference in the *Critique of Pure Reason*, according to which planetary gravitational perturbations do indeed attempt to explain all variations and apparent deviations of a unitary law (universal gravitation).[12]

ilean property (constant *g*) holds (to a high degree of approximation) only for falling bodies sufficiently close to the surface of the earth; for any considerable distance from the surface it must be generalized via the inverse-square proportion to the acceleration-field property.

11 See note 62 above. See also Smith (2012).

12 "[W]e arrive at unity of the genera of these paths according to their form [circular, elliptic, parabolic, and eventually hyperbolic orbits]; and we thereby further arrive, however, at unity of the cause of all the laws of their motion (i. e., gravitation). From there we afterwards extend our conquests further, seeking also to explain all variations and apparent deviations from these rules from the same principle" (Kant 1900–, A663/B691, my translation).

3 Einstein and the General Theory of Relativity

The Newtonian explanation breaks down, however, in the case of the perihelion of Mercury; it is then replaced by the Einsteinian explanation, which essentially involves a non-Euclidean spatial geometry. Here the Einsteinian explanation uses a new theory of gravitation, one which uses spatio-temporal curvature to represent the gravitational field. And an essential part of the Maxwellian *electromagnetic* field is its entanglement with spatio-temporal curvature as representative of the *gravitational* field. The electromagnetic field is an abstract measurable magnitude, like Newtonian gravitational force. Unlike the latter, however, it is also "*dynamical*"—and, in particular, propagated through space with a definite finite velocity *c*. It is thus an "actor" in the physical world along with the matter upon which it acts.[13]

The story of electromagnetism began with Hans Christian Oersted (1777–1851). Michael Faraday (1791–1867) then introduced the modern notion of "field" via induced currents. And Heinrich Hertz's (1857–1894) experiments with radio waves securely established the "structural" properties of the electromagnetic field via theory-mediated measurements, just as the corresponding properties of gravitational force were similarly established by Newton's "deductions from the phenomena". Hertz thereby established, in particular, the Maxwellian displacement current (see again Stein 1970).

The nineteenth-century evolution of electrodynamics as a whole was especially important. Early reactions to Oersted in the 1920s led to Ampère's "forces exerted by electricity in motion". In the 1840s Weber's "mechanical" action-at-a-distance approach in Germany was prominent. Helmholtz adopts it in the first edition of "The Conservation of Force" (1847). In the period from 1850 to 1880 Maxwell's elaboration of Faraday's field-theoretic approach became known on the continent. Helmholtz reacts in the second edition of his monograph (1881) with the formulation of his "neutral" action-potential approach. In 1884 Hertz, following Helmholtz, attempts to show that field-theoretic and "mechanical" approaches are equivalent. But he soon encounters insurmountable obstacles and realizes that only Maxwell can provide continuous (and finite) propagation. In the years 1884–88 Hertz attempts to demonstrate the existence of propagation, and he finally succeeds in this in 1888. Thus, the "long" nineteenth-century evolution of electrodynamics as

13 See Stein (1970).

a whole allows us to understand how the (temporary) end-point, Maxwell's electrodynamics, finally reaches a theoretical and experimental result.[14]

Maxwellian electrodynamics, however, is incompatible with Newtonian physics. The light-cone is *not* Galilean invariant, and the "structural" properties of Maxwellian causality are in fundamental tension with those of Newtonian (instantaneously propagated) causality. By contrast, space-time itself is Lorentz invariant. The "structural" properties of Maxwellian (finitely propagated) causality are also those of the (new) spatio-temporal background structure (Minkowski space-time).[15] Interferometer experiments of Michelson, Morley et al. in the early years of the twentieth century established these "structural" properties of the spatio-temporal background structure in the same sense that the corresponding properties of the electromagnetic field were established by Hertz.[16]

There was now a fundamental need for a new theory of gravitation, and general relativistic space-time filled the bill. Symmetric light-cones for the black-hole space-time, for example, are "vertical" in the distant external region, become tangent to the horizon, and then, in the internal region, lean towards the singularity. It is intended that the figure have the symmetries of translation up and down, and rotation about the axis.[17] And, more generally, the resulting four-dimensional space-time structure has implications for three-dimensional, purely spatial geometry as well. Indeed, one can visualize the purely spatial geometry in two dimensions by imagining that the Sun is a massive ball placed on a rubber sheet. The resulting geometry is circularly symmetric with increasing curvature as one radially approaches the central Sun, and this allows one to visualize Einstein's prediction of the deflection of light in the gravitational field of the Sun that was tested during eclipse observations in 1919.[18]

To fully appreciate the import of these observations, however, one needs to appreciate that the Einsteinian prediction was being tested against a corresponding Newtonian prediction (going back to the turn of the nineteenth century), whereby light was represented by a very small (and very light) massive particle moving

14 Two important works in this connection are J. Z. Buchwald (1994), *The Creation of Scientific Effects: Heinrich Hertz and Electric Waves* and O. Darrigol (2000), *Electrodynamics from Ampère to Einstein.*

15 Intuitive diagrams depicting the crucial differences between Newtonian and Minkowski space-time can be found in Friedman (1983). See also the beautifully intuitive treatment in Geroch (1978).

16 A clear and fairly detailed discussion of the Michelson-Morley (and related) experiments can be found in Resnick (1968). For a more historically oriented discussion, including the Lorentz-Fitzgerald contraction hypothesis and other such "compensatory" hypotheses, see Darrigol (2000).

17 See Geroch (1978). The light-cone diagram on the cover depicts the space-time structure within and around a black hole.

18 See the diagram in Misner, Thorne, and Wheeler (1973, 32).

at the (already known) velocity *c*. But the Newtonian predicted deflection turned out to be one-half the value of Einstein's. Thus, the observations and photographic measurements of the full Einsteinian deflection in 1919 empirically established the mathematical relationship between Newtonian and Einsteinian gravity (where gravity is represented as a curvature imparted to Minkowski space-time rather than as a Newtonian impressed force), just as the previous observations and experiments of Newton, Hertz, Michelson-Morley et al. established the mathematical properties of the different forms of causal interaction in these cases.[19]

The continuity in method and resulting epistemic status involved in all three cases captures the sense in which Kantian causal necessity can indeed be extended into the post-Newtonian period. Of course the nature of the causal relation undergoes important changes. In the Newtonian case the relevant force—the Newtonian gravitational field—mediates the interactions between material substances without being an actor in turn. The only substances, properly speaking, are massive material bodies, and they do not exchange momentum, in turn, with the (Newtonian) gravitational field. This situation fundamentally changes with the electromagnetic field, which is continuously propagated with a finite velocity rather than instantaneously; it thereby exchanges both momentum and energy with the relevant (charged) bodies on which it acts and is thereby acted upon. The situation changes even more radically in the general theory of relativity, where the gravitational field is not only propagated continuously but is now a field of geometrical curvature in the fabric of space-time itself. And this transformation of geometry—including even the three-dimensional geometry of space—into a physically acting and interacting field of curvature implies that Kant's conception of space as the a priori form of outer intuition has most definitely broken down. Nevertheless, a central element in the Kantian conception of causal necessity—its more than inductive and more than merely hypothetical epistemic status—is in fact preserved, relatively straightforwardly, even here.

The fate of Kant's conception of pure intuition is less straightforward, but there is a natural way to generalize his conception of space as an a priori form of intuition so as to embrace all the post-Newtonian developments we have considered. Instead of viewing this form as reaching globally into the cosmos—so as thereby to enable Newton's Euclidean treatment of the motions of the heavenly bodies within the solar system (and, for Kant, throughout the rest of the cosmos as well)—we view it as reaching only locally into a very small region (measured

19 See Kennefick (2009) for an excellent discussion of the 1919 observations. The advance of the perihelion of Mercury and other classic tests of general relativity can be treated similarly, albeit not so simply and intuitively. For details see Misner, Thorne, and Wheeler (1973).

in cubic metres) embracing the laboratory and/or observatory spaces of the relevant experiments. We see this, in fact, in all of our post-Newtonian cases: Hertz's experiments with radio waves, the Michelson, Morley et al., interferometer experiments, and the photographic observations and measurements verifying Einstein's prediction for the deflection of light.[20] In the last case, in particular, the space of the observatory is so small relative to the corresponding space of the cosmos (here extending well beyond the solar system) that it is, for all intents and purposes, infinitesimally small; so its geometry, even according to the general theory of relativity, remains Euclidean.[21] In this sense, it is still reasonable to view the use of Euclidean geometry as an a priori *constitutive* presupposition of the empirical observations that serve as tests for the theory, even though the global (cosmic) space employed by the theory is measurably non-Euclidean.

It turns out, however, that there is a serious problem with this line of reasoning. The relevant observatory—Harvard's Jefferson laboratory—generates relativistic curvature above the surface of the earth, where the laboratory in question still has the dimensions (approximately 22.5 metres) of the observatory at Sobral (which has only *negligible* general relativistic curvature). Thus, the Pound-Rebka experiment at the laboratory (1959) established the gravitational redshift by the emission and absorption of photons over this same distance (approximately 22.5 metres above the surface). This generates a successful application of the old quantum theory to general relativity, one which could not have been understood previously in the small region above the surface. Can Kant follow us here?

References

Buchwald, J. Z. 1994. *The Creation of Scientific Effects: Heinrich Hertz and Electric Waves.* Chicago: University of Chicago Press.
Cohen, I. B. 1980. *The Newtonian Revolution.* Cambridge: Cambridge University Press.
Darrigol, O. 2000. *Electrodynamics from Ampère to Einstein.* Oxford: Oxford University Press.
Friedman, M. 1983. *Foundations of Space-Time Theories.* Princeton: Princeton University Press.
Geroch, R. 1978. *General Relativity from A to B.* Chicago and London: University of Chicago Press.
Kant, I. 1900–. *Kants Gesammelte Schriften.* Edited by the Königlich Preußische, later Deutsche, then Berlin-Brandenburgische Akademie der Wissenschaften. Berlin: Reimer, later De Gruyter.

20 See Kennefick (2009, 41).

21 For example, the 1919 observatory at Sobral in Brazil was less than 10 cubic metres in volume; the stars observed belong to the Hyades cluster, the centre of which is approximately 150 light-years from the earth. In general, in any Riemannian manifold of any curvature and dimension, geometry in the infinitesimally small (of the tangent space at any point) is nonetheless flat or Euclidean.

Kant, I. 2004 [1786]. *Metaphysical Foundations of Natural Science.* Ed. M. Friedman. Cambridge: Cambridge University Press.

Kennefick, D. 2009. Testing Relativity from the 1919 Eclipse—A Question of Bias. *Physics Today* 62(3): 37–42.

Misner C. W., Thorne, K. S. and Wheeler, J. A. 1973. *Gravitation*, San Francisco: W. H. Freeman.

Newton, I. 1999. *The Principia: Mathematical Principles of Natural Philosophy*, 3rd edition (1726). Newly translated by I. Bernard Cohen and Anne Whitman. Berkeley: The University of California Press.

Resnick, R. 1968. *Introduction to Special Relativity.* New York, Chichester: J. Wiley.

Smith, G. E. 2002. The Methodology of the *Principia.* In: *The Cambridge Companion to Newton*, eds. I. B. Cohen and G. E. Smith, 138–173. Cambridge: Cambridge University Press.

Smith, G. E. 2012. How Newton's *Principia* Changed Physics. In: *Interpreting Newton*, eds. A. Janiak et al., 360–395. Cambridge: Cambridge University Press.

Stein, H. 1967. Newtonian Space-Time. *Texas Quarterly* 10(3), 174–200.

Stein, Howard. 1970. On the Notion of Field in Newton, Maxwell, and Beyond. In: *Historical and Philosophical Perspectives of Science*, ed. R. H. Stuewer, 264–287. Minneapolis: University of Minnesota Press.

Stein, H. 1989. Yes, But..., Some Skeptical Remarks on Realism and Anti-Realism. *Dialectica* 43(1–2): 47–65.

Worrall, J. 1989. Structural Realism: The Best of Both Worlds? *Dialectica* 43(1–2): 99–124.

Part III: **Epistemic Stances and Voluntarism**

Matthias Egg

Stances and Doctrines in Scientific Metaphysics

Abstract: The insight that philosophical positions ought to be understood as stances rather than doctrines is one of the cornerstones of Bas van Fraassen's critique of metaphysics. It has been argued, however, that not all kinds of metaphysics are equally vulnerable to this critique. In particular, James Ladyman and Don Ross have integrated van Fraassen's emphasis on stances into the defence of what they call *scientific metaphysics*. This raises the question to what extent such a metaphysics can still claim (as scientific metaphysicians routinely do) to be concerned with objective truth, which is usually understood as pertaining to doctrines, not to stances. It also transforms the way in which such a metaphysics might be defended against alternative philosophical projects, based on alternative stances. I will analyze this transformation in the context of Anjan Chakravartty's recent claim that no rational stance can be demonstrated to be superior to any other.

1 Introduction

Criticizing metaphysics is a tricky business, as one is in constant danger of reifying the ideas one employs in the critique, thereby in turn lapsing into a kind of metaphysics. In the course of introducing constructive empiricism, Bas van Fraassen has drawn attention to a famous instance of such a lapse, which occurred when earlier empiricists insisted that all our knowledge is based on *sense-data:*

> Such events as experiences, and such entities as sense-data, when they are not already understood in the framework of observable phenomena ordinarily recognized, are theoretical entities. They are, what is worse, the theoretical entities of an armchair psychology that cannot even rightfully claim to be scientific. I wish merely to be agnostic about the existence of the unobservable aspects of the world described by science—but sense-data, I am sure, do not exist. (van Fraassen 1980, 72)

More recently, van Fraassen (2002) has given a deeper and more general analysis of what exactly went wrong in the traditional empiricist critique of metaphysics, and indeed in other philosophical traditions as well. The basic misunderstanding, van Fraassen argues, consists in adopting the following (meta-)philosophical principle:

https://doi.org/10.1515/9783111019802-012

> Principle Zero:
> For each philosophical position X there exists a statement X+ such that to have (or take) position X is to believe (or decide to believe) that X+. (van Fraassen 2002, 41)

In the case of empiricism, X+ would be something like “Experience is the one and only source of information”. Adopting Principle Zero thus leads to a naïve kind of empiricism, which, as van Fraassen (2002, 42–46) shows, is subject to a *reductio ad absurdum:* Since empiricism is also committed to the principle that no factual claim can be ruled out *a priori*, it follows that, insofar as X+ is understood as a factual claim, metaphysical claims contradicting it are also factual and therefore cannot be ruled out *a priori*. But since metaphysical claims are not empirically testable, they cannot be ruled out *a posteriori* either. Hence, naïve empiricism can no longer serve as a basis for the task it was meant to perform: the critique of metaphysics.

Van Fraassen therefore rejects Principle Zero:

> A philosophical position can consist in something other than a belief in what the world is like. [...] [It] can consist in a stance (attitude, commitment, approach, a cluster of such—possibly including some propositional attitudes such as beliefs as well). (van Fraassen 2002, 47–48)

In order to escape the above *reductio*, he suggests to no longer understand empiricism as a *doctrine* (“a belief in what the world is like”, van Fraassen 2022, 47), but as a *stance*, characterized by

> (a) a rejection of demands for explanation at certain crucial points and
> (b) a strong dissatisfaction with explanations (even if called for) that proceed by postulation. (van Fraassen 2002, 37)

Furthermore, the empiricist (or empirical) stance includes the following principle that echoes the above-mentioned ban on ruling out *a priori* any factual claim:

> (c) As in science, so in philosophy: disagreement with any admissible factual hypothesis is admissible. (van Fraassen 2002, 43)

The potential for renewal held by the shift from doctrines to stances is, however, not limited to a reconceptualization of empiricism. Other philosophical positions have as well suffered from an attachment to Principle Zero, and it is useful for what follows to look at the second example that van Fraassen discusses in this context: materialism. At first sight, naïve materialism, characterized by a statement X+ like “Matter is all there is” (van Fraassen 2002, 49), may not be subject to a similar problem as naïve empiricism, because materialism (unlike empiricism) is not obliged to eschew metaphysics. But the problematic nature of X+ becomes apparent as soon as one asks what is to count as "matter". If the content of the term "matter" is to be specified by current science, then materialism is likely to be falsified by future scientific discoveries. If, on the other hand, the specification of "matter" is deferred to the end of scientific inquiry, then materialism is empty until this end is reached.

As in the case of empiricism, this problem can be avoided by moving from naïve materialism to the *materialist stance*, characterized by "a certain deference to the content of science" (van Fraassen 2002, 62). Notice that the contrast to the empiricist stance is not that the materialist stance involves a greater admiration for science. Empiricism, too, is very positive towards the empirical sciences, but "this admiring attitude is not directed so much to the content of the sciences as to their forms and practices of inquiry" (van Fraassen 2002, 63).

We will come back to the disagreement (and the possibility of a reconciliation) between the empiricist and the materialist stance in Section 5, but the main aim of this chapter is to assess the impact of the shift from doctrines to stances on the recent (and possible future) debate in metaphysics. For that purpose, I will first describe how the current trend known as *scientific metaphysics* is, at least by some of its proponents, understood as exemplifying that shift (Section 2), resulting in a distinction between *strong* and *weak* metaphysics. The significance of this distinction will be critically discussed in Section 3. In Section 4, I turn to the question how the nature of the conflict between different philosophical positions changes if these are no longer understood as doctrines but as stances. The concluding Section 5 applies these considerations to the stance that underlies the project of scientific metaphysics.

2 Scientific Metaphysics as "Weak" Metaphysics

Although, as we have seen, the notion of stance may be applied to better understand different philosophical positions, it is no accident that it is introduced in the context of empiricism, understood as "a recurrent rebellion against the metaphysicians" (van Fraassen 2002, 36). For even before one endorses the empirical (or any other anti-metaphysical) stance, there is an anti-metaphysical element in the very move from philosophical doctrines to stances, insofar as metaphysics is usually understood as making doctrinal claims. This requires that if a kind of metaphysics is to survive the move from doctrines to stances, it must consist in something other than non-empirical doctrinal beliefs about reality.

It is no surprise, then, to find a discussion of stances in one of the most influential recent attempts to develop a metaphysical position that does not fall prey to van Fraassen's critique and to other instances of the "recurrent rebellion".

The position is called "naturalized metaphysics" (Ladyman and Ross 2007) or, more recently, "scientific metaphysics" (Ross et al. 2013), and its authors share with van Fraassen an acute awareness of the shortcomings of traditional (and, to their dismay, partly ongoing) projects in metaphysics (see, in particular, the opening chapter of Ladyman and Ross 2007). One of their means of distancing themselves from such projects is the distinction between what they call "strong" and "weak" metaphysics. Traditional metaphysics, concerned as it is with "non-trivial doctrinal beliefs about the structure of the world that go beyond what the sciences tell us or imply" (2007, 60), falls into the former category, about which Ladyman and Ross are equally dismissive as van Fraassen:

> To agree with van Fraassen that philosophical positions ought to be stances rather than doctrines is to suppose that there is value in some philosophy, somehow conceived, but not in strong metaphysics. (Ladyman and Ross 2007, 60)

In contrast, the *weak* kind of metaphysics they propose as the type of philosophy that is still of value consists in articulating a unified world-view on the basis of a given stance. Their own stance, called "scientistic", will be more closely inspected in Section 5, but for the moment, I am concerned with the more general question how the notion of stance underwrites a workable distinction between strong and weak metaphysics. The idea seems to be that the propositions of weak metaphysics are not claimed to be true in an absolute sense, but only relative to a given stance:

> *Within* the context of our scientistic stance, we will defend various propositions that will have the grammatical form of saying what there is. These should be understood not as doctrines but as proposed provisional commitments for living out the stance. (Ladyman and Ross 2007, 65)

In the next section, I will raise some doubts about this alleged non-doctrinal character of weak metaphysics.

3 Is Weak Metaphysics Less Doctrinal?

According to the description just quoted, there are two features that distinguish the propositions of scientific metaphysics (understood as an instance of weak metaphysics in the above sense) from the doctrines of strong metaphysics: They are provisional and they are stance-relative. We will now see that both these features are undermined by further elements of scientific metaphysics, as defended by Ladyman and Ross.

One might think that, by calling their commitments "provisional", Ladyman and Ross are responding to the problem (mentioned in connection with materialism in Section 1 above) that current science can be expected to be overthrown by future science, such that metaphysical claims derived from current science are likely to be false. And indeed, they mention this problem immediately after introducing the notion of weak metaphysics. But in response to it, they explicitly deny "that the history of science is most persuasively presented as a history of revolutions in which theories are repeatedly routed by successors and found to be false". Instead, they think that there is "a defensible basis for joining the Logical Positivists in viewing the history of science as a history of progressive accumulation of knowledge" (2007, 65). Presumably, this basis is provided by the structuralism they develop in the subsequent chapter of their book (2007, Ch. 2). But if there really is a progressive accumulation of knowledge, this means that the well-established parts of current science are *not* provisional, hence it is unclear in what sense the commitments of scientific metaphysics are claimed to be provisional.

The second feature, stance-relativity, is more interesting. It promises to set weak metaphysics apart from strong metaphysics by acknowledging that there is no such thing as a metaphysical truth that holds independently of the stance within which it is formulated. The best way to see what this amounts to is to look at the way in which Ladyman and Ross propose to engage with people who resist adopting the scientistic stance:

> Their resistance to science, which must be quite thoroughgoing if it is not to be unprincipled, will confront them with serious policy problems in the management of social affairs, and we will want to press them as hard as possible on these. But we would not try to convert them with metaphysics, for van Fraassen is right that that would require strong metaphysics, and strong metaphysics can't get off the ground. (Ladyman and Ross 2007, 64)

If this were correct, it would mean that weak metaphysics differs from strong metaphysics by not triggering any temptation to convert someone from one stance to another. However, Ladyman and Ross do not seem to be immune to that temptation:

> The best motivation for trying to synthesize our scientific knowledge into a unified picture—that is, for building naturalistic metaphysics—is the crucial service this activity potentially performs in extending the Enlightenment project. If science is not seen to provide the basis for a general worldview, then people will continue to collectively confabulate alternative general pictures. This in turn matters because the confabulated pictures inspire groundless and usually wasteful and destructive politics and policy. (Ladyman and Ross 2013, 113)

Obviously, the kind of metaphysics described here is *weak*, being simply the articulation of a coherent world-view on the basis of the scientistic stance. Nevertheless, the passage implies that this activity has the potential of dissuading people from confabulating alternative (that is, non-scientistic) general pictures, and it is hard to see how this differs from "trying to convert them with metaphysics".

A final piece of evidence against the non-doctrinal character of weak metaphysics is the non-equivalence of different stances with respect to objective truth. Again, this appears most clearly in a passage that contrasts the scientistic with a non-scientistic stance:

> We can say 'go in peace' to Heideggerians, noting that it was entirely appropriate that Heidegger did not attempt to base any elements of his philosophy on science, and focused on hammers [...] rather than atoms [...] when he reflected on objects. We, however, are interested in objective truth rather than philosophical anthropology. Our quarrel will be with philosophers who claim to share this interest, but then fail properly to pay attention to our basic source of information about objective reality. (Ladyman and Ross 2007, 5)

Here, the adherents of the scientistic stance do indeed refrain from trying to convert those who inhabit a different stance, but this is only because the latter is not viewed as a genuine competitor when it comes to the pursuit of objective truth. The underlying assumption seems to be that one can only dismiss the scientistic stance insofar as one is not interested in objective truth. After all, science is "our basic source of information about objective reality". Such a view, while honouring the letter of weak metaphysics, violates its non-doctrinal spirit: One may well declare that the statements of scientific metaphysics are just "proposed commitments for living out the [scientistic] stance" (Ladyman and Ross 2007, 65), but since that stance is singled out as being concerned with objective truth in a way in which other stances are not, these commitments again start to look a lot like the doctrines of strong metaphysics.

All this suggests that the distinction between strong and weak metaphysics is rather unclear and therefore unlikely to constitute a significant advantage of scientific metaphysics over other metaphysical projects. Rather, the lesson to be drawn from van Fraassen's rejection of Principle Zero is that a defence of scientific metaphysics must include giving reasons for adopting the scientistic stance. However, the recent debate on stances has seen some doubts about the very possibility of rationally defending the adoption of a certain stance. This is the topic of the next section.

4 The Difficulty of Evaluating Stances

A major theme in van Fraassen's (2002, Lecture 3) discussion of stances is the difficulty of understanding "revolutions/conversions" in our thinking that were so radical that the posterior position seemed absurd from the perspective of the prior one, rendering it impossible to rationally justify the transition between them. The general problem is that each stance comes with its own set of values, whereas justifying any preference for one stance over the other would seem to presuppose an external standard or measure applicable to both of them. This does not only hold for the diachronic case of replacing one stance by another, but also for the synchronic comparison of alternative stances, an issue recently discussed by Anjan Chakravartty (2017). Although Chakravartty does not describe the situation as one of complete incommensurability, he views the only common measure as an insufficient basis for a ranking of stances:

> Since rationality is the only stance-neutral criterion for the acceptability of a stance, there are no further grounds on which to prosecute a non-question-begging case for the epistemic superiority of one over another; they are, *qua* rationality, the only relevant measure, 'equally strong'. (Chakravartty 2017, 243)

Looking at the two examples of stances that figure most prominently in Chakravartty's discussion (the first one being the familiar empiricist stance, the second one the "metaphysical stance" (Chakravartty 2017, 212) defined roughly by negation of the empiricist policies (a) and (b) from Section 1), one can certainly agree that they both count as rational. Hence at this coarse-grained level, it would indeed be wrong to declare either of the two stances superior to the other. Chakravartty, however, argues for a more general thesis which I want to question in what follows: "Regarding the question of superiority [of any given rational stance] there can be only *aphasia*, speechlessness, and suspension of judgment" (Chakravartty 2017, 244).

The reason why I think such a conclusion is premature and unduly defeatist is that the stances we encounter in actual debates on metaphysics are never purely empiricist nor purely metaphysical. On the one hand, pursuing a metaphysics that floats free from any constraints imposed by empirical findings would amount to a kind of dogmatic slumber that is simply unavailable to any philosopher since Hume. On the other hand, Chakravartty (2017, Ch.2) himself makes a convincing case that a certain amount of metaphysics is inevitable in our reflecting on science, even if we go about it with a firmly empiricist outlook. We will encounter further reasons for mixing empiricism with metaphysics in the next section, but let me here focus on the impact this mixing has on the rationality of stances. Chakravartty explicates rationality in terms of internal coherence (2017, 224) and then defends this conception against the worry that it might be too permissive (2017, 231–234). If the mere avoidance of self-contradiction is all it takes to count as rational, this worry goes, and if all rational stances (in this generous sense) are acceptable, doesn't this force us to accept all kinds of pathological stances, as long as they do not induce their adherents to contradict themselves? Chakravartty responds to this by emphasizing that internal coherence does not only have a logical but also a pragmatic dimension, which inhabitants of pathological stances are liable to violate, resulting in a "self-sabotage by their own lights" (2017, 224). I think this response is on the right track and I agree that "pragmatic coherence is a richer notion than has been appreciated" (2017, 233), but it seems to me that the resulting enrichment of the notion of rationality not only has the consequence of denying rationality to the pathological stances, but also of opening the possibility that some non-pathological stances satisfy the requirement of rationality to a larger degree than others. And in light of the above, this possibility seems to be realized, insofar as any liveable stance in philosophy of science inevitably involves a mixture of mutually conflicting (empiricist and metaphysical) elements. In many cases, the internal conflict may not be stark enough to render the stance irrational. As Chakravartty puts it: "it can be extremely difficult to demonstrate that a stance is internally incoherent" (2017, 225). But there is a conflict nonetheless, and the assessment of its severity is a way of evaluating the stance that does not itself depend on any stance-relative values. I will finish this chapter by sketching such an evaluation for a concrete case.

5 Elements of a Stance Evaluation: The Case of Scientism

A useful way for getting a better grasp of Ladyman and Ross's scientistic stance is to ask why one cannot rest content with a purely empiricist stance. Doesn't empiricism's admiring attitude towards the forms and practices of scientific inquiry mentioned in Section 1 sufficiently capture the responsiveness to science that should be required from a reasonable philosophical position? Similarly to Chakravartty (2017, Ch. 2), Ladyman (2011) thinks that it does not. More specifically, he argues that empiricism by itself can accommodate neither the role played by metaphysics as a source of inspiration for further empirical research, nor the modal content of scientific theories. These are grounds for doing metaphysics stemming from the sciences themselves.

Another motivation for moving beyond the empirical stance was already mentioned in Section 3 above: The empiricist, in line with his rejection of demands for explanation and his dissatisfaction with explanatory postulates, is also suspicious of any attempt to synthesize our scientific knowledge into a unified (metaphysical) picture that would be capable of directing people away from confabulated alternative pictures and destructive policy. The scientistic stance thus differs from empiricism by attributing to science a life-orienting role, mediated by a properly scientific metaphysics.

It is worth noting that van Fraassen himself implicitly acknowledges the limits of empiricism in this respect. Empiricism does not provide us with a sufficient account of the world and our place in it, because, for example, the question which of the beings around us are to count as persons is not an empirical one (van Fraassen 2002, 189–193). What van Fraassen denies is that any metaphysics ought to be invoked in order to overcome this limitation of empiricism. His own response to it (2002, Lecture 5) refers to God, but this God is emphatically *not* to be confused with the metaphysical construct that many philosophers discuss using that name (see the opening remarks of Lecture 1, where van Fraassen quotes Pascal's phrase "The God of Abraham, Isaac and Jacob, not the God of the philosophers"). Nonetheless, it marks a clear departure from the purely empiricist stance and thereby further illustrates my doubts that anyone in the actual debate adheres to something like the unmixed stances which appear in Chakravartty's discussion.

The comparison between van Fraassen's empiricist-non-secularist and Ladyman's scientistic stance would thus be an interesting test case for Chakravartty's denial of non-question-begging superiority assessments for stances, but it is beyond the scope of this chapter. Instead, I will limit the rest of my discussion to the internal coherence of the scientistic stance, which constitutes the basis of scientific metaphysics. To what extent other stances score better (or worse) on internal coherence will have to be investigated in future work.

Ladyman and Ross (2007) and Ladyman (2011) describe the scientistic stance as a combination of the empirical and the materialist stance. This is somewhat surprising, because van Fraassen had introduced these two stances as being in contrast with one another. To see that the contrast may be reconcilable, notice the difference between the materialist stance and what Chakravartty calls the metaphysical stance. The latter is indeed simply the opposite of the empirical stance, so the idea of combining the two would be absurd. Not so for the materialist stance, which, although being somewhat metaphysical, can also be seen as somehow orthogonal to the empirical stance. We saw in Section 1 that van Fraassen characterizes materialism as a certain deference to the content of science. He then goes on to say:

> This deference takes two forms: the belief that the scientific description of the world is true, in its entirety or near enough, and at least a strong inclination toward completeness claims for the content of certain sciences. (van Fraassen 2002, 63)

This certainly involves some metaphysics (at least on the prevalent understanding of truth in this part of the debate), but it does not obviously contradict any of the empiricist policies (a) to (c). Hence Ladyman and Ross, endorsing van Fraassen's descriptions of empiricism and materialism, declare:

> So we really are, in detail, both perfect empiricists and perfect materialists according to van Fraassen's criteria. Note that this cannot be diagnosed as implying doctrinal self-contradiction, for that is a charge that is inappropriate to stances. (Ladyman and Ross 2007, 63)

In response to this proposal, van Fraassen (2011, 167–168) briefly mentions some ways in which the scientistic stance can nevertheless fail to be internally coherent. I shall conclude this chapter by providing further illustration of how this possible failure may manifest itself when the scientistic stance is applied to a concrete scientific theory, namely quantum mechanics.

The reason why quantum mechanics is a particularly interesting case for a possible reconciliation between empiricism and materialism is a certain tension entailed by the so-called measurement problem.[1] Very briefly, the problem can be put as follows: Quantum mechanics provides some highly reliable rules for the (probabilistic) prediction of measurement outcomes, but it is notoriously unclear what counts as a "measurement" and how these rules of prediction relate to the way physical systems behave when no measurement is performed. There are various ways of responding to this problem, and a very natural way of classifying them reveals a dilemma for the scientistic stance: On the one hand, there are attempts to dissolve the problem by denying a representational role to the quantum-mechanical formalism. If quantum mechanics is nothing but a tool for managing our expectations about the outcomes of experiments, then the problems surrounding the status of "measurement" are no longer in need of a principled solution, but can be dealt with pragmatically. However, such a non-representational view is likely to conflict with the materialist stance and its insistence on the truth-content of scientific descriptions. On the other hand, responses to the measurement problem which seek to uphold a representational role of the theory involve the kind of metaphysical postulates that the empirical stance disallows; some of them introduce additional variables that do not change the empirical content of the theory, others modify the theory's dynamical laws without (measurably) altering its empirical predictions, and yet others postulate a multitude of empirically inaccessible parallel universes. In short, any approach to the measurement problem seems to conflict with either the materialist or the empiricist stance.

1 For an introduction to the problem and further references, see, e.g., Myrvold (2017). For more details on the following case study, see Egg (2019).

Being aware of the possibly disastrous consequence that this would have on the internal coherence of the scientistic stance, Ladyman and Ross (2007, 2013) seek to develop a response to the measurement problem that avoids unduly metaphysical postulates but still grants a sufficiently representational role to the quantum formalism. I have argued (in Egg 2019) that their attempt is unsuccessful, but this does not mean that no such attempt can succeed. Indeed, I take my own effort in that direction (Egg 2021) to show that one can attribute quite a lot of representational content to quantum mechanics without diving into metaphysical speculation. At the same time, however, it also clearly delimits the prospects of simultaneously satisfying empiricist and materialist demands. Without going into details, let me summarize the lesson from this case as follows: If there is (as the scientistic stance supposes) a kind of metaphysics that is compatible with the empirical stance, it must only involve postulates that are in some sense shared by the different approaches to the measurement problem (at least those that I classified above as "representational"), and it turns out that such a metaphysics can be neither fundamental nor complete. But this is to say that it does not fully respond to the requirements of the materialist stance.

This kind of consideration shows that for stances that are complex enough to do any real work in philosophy of science (such as the scientistic stance that underlies the influential project of scientific metaphysics), internal coherence is not as easy to get as Chakravartty supposes. This opens the door to more interesting stance-independent evaluations than might initially seem possible. And it therefore shows that the emphasis on stances instead of doctrines, for all its transformative effects on the debate between conflicting philosophical positions, need not land us in a state of *aphasia* with regard to superiority assessments among them.

References

Chakravartty, A. 2017. *Scientific Ontology: Integrating Naturalized Metaphysics and Voluntarist Epistemology.* Oxford: Oxford University Press.

Egg, M. 2019. Dissolving the Measurement Problem Is Not an Option for the Realist. *Studies in History and Philosophy of Modern Physics* 66: 62–68.

Egg, M. 2021. Quantum Ontology without Speculation. *European Journal for Philosophy of Science* 11(32): 1–26.

Ladyman, J. 2011. The Scientistic Stance: The Empirical and Materialist Stances Reconciled. *Synthese* 178: 87–98.

Ladyman, J. and Ross, D. 2007. *Every Thing Must Go: Metaphysics Naturalized.* Oxford: Oxford University Press.

Ladyman, J. and Ross, D. 2013. The World in the Data. In: *Scientific Metaphysics*, eds. D. Ross, J. Ladyman and H. Kincaid, 108–150. Oxford: Oxford University Press.

Myrvold, W. 2017. Philosophical Issues in Quantum Theory. In: *The Stanford Encyclopedia of Philosophy* (Spring 2017 Edition), ed. E. N. Zalta. https://plato.stanford.edu/archives/spr2017/entries/qt-issues/ (last accessed: 17. 9. 2022).

Ross, D., Ladyman, J. and Kincaid, H. (eds.). 2013. *Scientific Metaphysics.* Oxford: Oxford University Press.

van Fraassen, B. C. 1980. *The Scientific Image.* Oxford: Clarendon Press.

van Fraassen, B. C. 2002. *The Empirical Stance.* New Haven: Yale University Press.

van Fraassen, B. C. 2011. On Stance and Rationality. *Synthese* 178: 155–169.

Nancy Cartwright

Evidence, Relevance and Warrant: In Defence of Voluntarism

Abstract: Bas van Fraassen endorses a voluntarist epistemology. I do too, influenced by him and by Otto Neurath, though I confine my claims to the epistemology of *science*, where the bulk of my work is located. Here I present arguments for voluntarism additional to van Fraassen's own, based on accounts of what makes for scientific evidence. I take it that evidence claims for a hypothesis should be true and should genuinely speak for or against the hypothesis. A hypothesis is warranted for us only if we are warranted in taking each of these to be the case. Both, I argue, lead to voluntarism via the familiar route pointed to in Neurath's famous boat metaphor. There is no end to the warranting game. Warrant is always conditional on a myriad of claims we have already plumped for. The second requirement, relevance, is my chief focus here, since, I argue, warranting an evidence claim is a doddle compared with warranting that it *is* evidence. If we want evidence then, plumping is ubiquitous.

1 What This Paper Is About

Bas van Fraassen calls his epistemology "voluntarist". I have learned from him—and Otto Neurath—to be a voluntarist too. In this paper I will provide some reasons to be a voluntarist beyond van Fraassen's own reason.

Van Fraassen explains that his epistemology is voluntarist in that "it makes judgment [...] a matter of cognitive commitment, intention, *engagement.* Belief is a matter of the will" (van Fraassen 2008, 57). According to him,

> [E]pistemic judgments are most like expressions of intention [...] [What] is conveyed [is] that I have made a decision, have formed an intention, am committed to a certain stance or program or course of action. (van Fraassen 2008, 56)

Probably the most straightforward route to voluntarism is via Neurath's boat. Any empirical claim h needs evidence e in its support and for h to be warranted e must be warranted. So e needs evidence, warranted evidence; and so on. There is no way to end this ever-extending net of warrant but to plump. I fear we have gotten so used to this regress that we don't feel its bite, and some philosophers also think they have ways to avoid it.

https://doi.org/10.1515/9783111019802-013

Van Fraassen offers a different route to voluntarism in his work on Bayesian updating and incoherence. Here I offer yet another for those like me who never got on board with Bayesianism. Mine is similar to the familiar Neurath's boat route but one level up, where the problems multiply and in all directions at once. This has to do not with establishing evidence claims but with relevance: showing that these hard-won evidence claims have any bearing on the hypothesis they are meant to evidence. Warranting an evidence claim e is a doddle compared with warranting that e *is* evidence.

Throughout I shall distinguish between *evidence* and *warrant.* I take there to be a fact of the matter, independent of us, whether or not a fact e bears on the truth of some different (possible) empirical fact expressed in hypothesis h,[1] that is, whether e is *evidence for* h. I use *warrant for a claim* to refer to *our entitlement* to take the claim to be true.

My central argument looks like this:

i. No empirical claim is warranted without evidence.
ii. For entitlement, it is not enough to *have* evidence. There must be warrant that it *is* evidence.
iii. No fact e comes warranted that it is evidence for any hypothesis h (what is and what is not evidence for h is never 'given').
iv. So there must be evidence e' that e is evidence for h.
v. For h to be warranted, it must be warranted that e' is evidence that e is evidence for h.
vi. That warrant requires further evidence, which expands the evidence requirements far beyond what is needed to warrant simple empirical claims.
vii. There is no warranted stopping point.

So, plumping is unavoidable.

I am going to exemplify these claims with respect to two theories of evidence: the very familiar HD theory and the *argument theory of evidence* that I endorse, which turns HD on its head, plus some short remarks about probability-based theories of evidence. In Section 4 I will illustrate vi. with the example of warranting (or in this case failing to warrant) that the success of the Integrated Nutrition Programme (INP) in several sites in Tamil Nadu is evidence for INP's general success or at least its general success in the Indian sub-continent.

1 Henceforth I shall not struggle to make use/mention distinctions where no confusion is likely to arise.

2 Three Requirements

I propose three requirements on warrant. To warrant a claim h, we need to have
a. *Sound* evidence for h
that
b. *speaks for* h
plus
c. *warrant* that a. and b. hold.

In combination with c., I claim that a. and b. both separately end us in voluntarism. The first via the usual route of Neurath's boat. The second, because in all cases of warrant

We need evidence that our evidence is evidence.

And, if h is to be warranted, this evidence in turn needs to be evidenced and correlatively we need evidence that the evidence for our evidence *is* evidence for it. And so forth. We only end the chain of regress by stopping at some point, just accepting some body of claims, at least for the nonce—that is, by plumping. We may choose some system for deciding—flip a coin, accept what the vicar says, accept what Jane Goodall or Sunetra Gupta claims, accept almost of all that we have inherited (albeit even without recognizing what that is), accept the evidence of our senses as we conceive it and so on. But in the end that is a choice *we* make. In van Fraassen's words: we have "made a decision", "formed an intention", "committed" ourselves to a course of action.

Consider *sound evidence* first. Here the argument is familiar, from the failure of 'the given' to be given. We want our evidence claims, e, for a hypothesis to be true, and, since we are in the business of warrant, we want there to be good reasons to count them as true. But good reasons for e will include evidence for e. So we are off and running on the regress just noted. We'd love to end this regress in something that just comes warranted and doesn't earn its warrant from elsewhere. But there is no such thing, a lesson encoded in Otto Neurath's famous boat metaphor:

> There is no way to establish fully secured, neat protocol statements as starting points of the sciences. There is no *tabula rasa.* We are like sailors who have to rebuild their ship on the open sea, without ever being able to dismantle it in dry-dock and reconstruct it from its best components. (Neurath 1983, 92)

Now turn to *relevance*, which is my central topic here. There is no fact of the matter built into a fact by itself that fixes what it can support or discredit. In Jane Austen's *Sense and Sensibility*, we are told of Marianne's sister Elinor that

> in his addressing her sister by her christian name alone, she instantly saw an intimacy so decided, a meaning so direct, as marked a perfect agreement between them. From that moment she doubted not of their being engaged to each other. (Austen 2006 [1811], Ch. 12, 70)

Why is it that Willoughby's use of Marianne's first name is evidence for her sister of their engagement? Because of the mores of their social setting and the assumption that Willoughby is a gentleman. So the answer depends on a great many other facts. This is always the case.

For a more scientific example, consider Maria Elena Di Bucchianico's (2009) story of two warring camps in high temperature superconductivity. One took the explanatory mechanism to be phonons; the other, magnetic modes. In 2002, new methods rigorously showed a 'kink' in the dispersion curve of reflected photoelectrons. Both camps agreed that the data were correct. But they had wildly different interpretations of it due to the great number of differing assumptions they were also committed to. Each of the two warring camps claimed this evidence supported their theory and was incompatible with the opposition's.

So, no matter how secure we are in the knowledge of a fact, that is not enough to know what it is warrant for and what not. No matter what e and h are, whether e is or is not evidence for h depends on what other facts hold; and to be warranted in taking e as evidence for or against h we must have warrant for these facts and also for the claim that they do indeed support or discredit h. Those claims themselves then need evidence to support them, and so on. We are once again off and running on an infinite regress that has no self-warranting stopping place.

So the failure of 'the given' to be given never stops biting. Maybe you think that there's a way to get around the first set of Neurath's boat problems: there is a way to establish our evidence claims well enough that you are rationally entitled to take them as true, with no further plumping necessary. Still the problem of establishing relevance looms. Warranting that these claims *are* evidence and *for what* takes plumping. There are two roads to plumping via relevance. One is via the usual evidence regress; another is warranting that we have enough of the right sorts of reasons. The rest of this paper will expand on these.

3 The Argument Theory of Evidence

What makes a fact e relevant to/bear on/speak for(against) hypothesis h? The lesson to be drawn from the discussion in Section 2 is that evidence is not a 2-place relation 'e is evidence for h'. Rather,

> *Evidence is a 3-place relation.*

What is the third place? What is it that secures the *relevance* of e to h?

The issue of evidential relevance has been on the tapis in contemporary philosophy of science since at least C. G. Hempel's famous hypothetico-deductive (HD) model (Hempel 1965), in which h explains e if e can be deduced from h (plus some auxiliary assumptions) and correlatively, in that case e confirms h (to some extent) —which implies e's evidential relevance to h. Looming in the background, however, is the spoiler: Karl Popper's insistence that the HD model commits the fallacy of affirming the consequent. As we know, Popper argued that there is no warrantable theory of induction, there is only deduction. That means that there is really no way to confirm hypotheses, only to disconfirm them and correlatively HD can supply no basis for evidential relevance except for facts that are incompatible with the target hypothesis.

I rehearse these well-known discussions because I am keen on HD, but I want turn HD on its head. In HD we start with h; then see what it takes to get to e. I propose that we start with e; then see what it takes to get to h. Think: what kind of a case in which e plays a role can secure confidence in h? Then e is evidence for h relative to such a case. Why? Because if you can't make a case for h in which e matters, you've no reason to claim e bears on the truth of h.

So: I claim that the third place is to be filled by a *good argument*, a view, called 'the argument theory of evidence', that I share with others including Michael Scriven (1966, 1976, 1981, 1998) and Julian Reiss (2015). This theory maintains that

> e is evidence for h *relative to a good argument from e to h.*

The argument shows e's relevance to h. Recalling the distinction between evidence and warrant, there must be a good argument for e to be evidence and for e to figure in warranting h requires that there be warrant that there is such an argument.

Think again about Elinor's conclusion in *Sense and Sensibility.* We suppose that it is supported by the following argument, although probably not consciously formulated by Elinor:

R: A gentleman will only call a woman outside the family of the same social status by her first name if they are engaged or married.
WG: Willoughby is a gentleman.
FN: Willoughby used Marianne's first name.
Therefore: W & M are engaged.

Elinor's premises R and WG are what makes FN evidence that Marianne and Willoughby are engaged.

This, I claim, holds in general.

AT: e is evidence for h *relative to* a good argument A for h,

where *good* arguments are valid and sound.[2] So we should not talk about e being evidence for h but should always have in view an argument that fills in the third place. One great advantage of the argument theory is that it makes you concentrate on just what it is you are assuming when you take some facts to evidence h. It is like John Norton's material theory of induction in which "inductive inferences are not warranted formally but materially. They are justified by facts, that is, by the factual matter that is itself the content of argumentation" (Norton 2014, 2). There are more than 2000 swans in the Thames. They are all white. In the right season, there are more than 40,000 swans in the UK. They are all white. We may be tempted to conclude 'All swans are white'. But what would make these 40,000 swans evidence that all swans are white? The truth of 'All swans have the same colour' would do. With this, there is a valid argument for the hypothesis—though an argument that does not secure the relevance of the 40,000 white swans because, as we see in Sidney Harbour, all swans do not have the same colour.[3] So, I claim, if you aren't doing arguments, you aren't doing evidence, because evidence is relative to the facts that make up good arguments.

The argument theory of evidence takes us straight to voluntarism via an evidence regress. Recall,

E is not warranted as evidence for h unless there's evidence e is evidence for h.

2 My requirement that the argument be valid may be thought too strong. Reiss's theory of evidence, for example, requires only a good argument, as does John Norton's material theory of induction. The difference does not matter significantly for issues about voluntarism.

3 I claim the 40,000 swans are not evidence. This is independent of whether we might have, or not have, good reason to suppose they are evidence, as for example in having good reasons to think all swans have same colour.

To warrant h we need some evidence claims e_h that together with a set of additional assumptions $\{A_{h1}, ..., A_{hn}\}$ argue for h. If this argument is sound each A_{hi} should be true. And if h is to be warranted, each A_{hi} should be warranted, so it must have evidence e_{hi} in its favour. This means that e_{hi} along with a set of additional premises $\{A_{hi1}, ..., A_{hin}\}$ imply A_{hi}, which premises should be true and, if A_{hi} and thence h is to be warranted, each A_{hij} must be warranted and so must have evidence in its favour. And so on and so on.

This leaves us ultimately with only the possibility for conditional warrant. *Given* the premises for a valid argument for h, h is warranted for us. We are left, as van Fraassen says, to 'decide' and 'commit' to the premises if we are to have any kind of warrant at all.

Given this conclusion, perhaps in an effort to avoid voluntarism we should be more reluctant to embrace the argument theory of evidence that I endorse. Perhaps we should revert to one of the more conventional accounts of confirmation —like HD. Let's consider it then.

4 Why HD Doesn't Help, Nor Probabilistic Accounts Either

The HD account looks at very first blush as if it is a purely formal account of relevance: e is relevant to h if e can be deduced from h. But we know that isn't so, for two reasons. First, for the most part the evidence claims we want to use to support a hypothesis h do not follow from h alone but from h plus other assumptions, usually called 'auxiliaries'. I take it that it is clear how this lands us in a regress as soon as we demand that no hypothesis h is warranted by a fact e unless the auxiliaries that help deduce e are warranted. Second, we are not just interested in having some facts that speak for h however weak their voice. We want enough of the right kinds of facts to *warrant* h. Philosophers of science have supplied a canonical list of requirements to add to HD to do the job, what I call 'HD with bells on'—*HDB*.

Different philosophers offer different requirements. All agree that just deducing a lot of facts from h is radically insufficient for warranting h. I will illustrate with a set of requirements that are widely shared:

HD warrant for h requires that there be a set E(h) of e's implied by h that are:

1. True
2. Numerous
3. Varied

4. Novel
5. Differentiating between h and alternatives

It is not enough for warrant, though, that there *be* enough of the right kind and variety of (true) facts implied. We should have warrant for this ... reasons that support it. What backs up the claim that, though not all the e_i's we adduce as evidence need be true, a sufficient number of the right kinds are? We need warrant for the claim that *enough* are true, that *enough novel* ones are true, that the set of true ones has enough of the *right kinds of variation* and that *enough true ones genuinely differentiate* h from alternatives. Let's call a subset of e_i's from among those on offer for which we have warrant that it satisfies all these requirements with respect to hypothesis h a *warranting set for h – W(h).* I will look at variation as an example of how these requirements support my claim that voluntarism is inevitable.

What kind of variation is needed and what reasons can be adduced in favour of, first, the claim that that is the right kind of variation, and second, that a large enough set of true e_i's exhibit this kind of variation? As usual, the answer differs case by case. Let us look at one in particular involving the Integrated Nutrition Programme (INP), which was proposed by the World Bank and initiated in Bangladesh in 1995.[4]

The hypothesis in question is this: *In places like Bangladesh where "behaviours related to feeding of young children have at least as much (if not more) to do with the serious problem of malnutrition ... as poverty and the resultant household food insecurity do"* (White and Massett 2007), *the Bangladesh Integrated Nutrition Programme (BINP) will improve infant nutrition.* The main activity of BINP is to educate mothers about how to use resources available to provide better nutrition for their children. This hypothesis turns out not to be true despite good reasons in its favour. In particular it failed in Bangladesh.[5]

The good reasons were indeed varied, going even beyond HD-type evidence, including for example both a general theory and a causal model of how the effect is supposed to be produced. As Howard White and Eduard Massett report in their post hoc evaluation of the Bangladesh failure, the basic theory is simple: where 'bad practices' are responsible for malnutrition 'changing bad practice to good will bring about nutritional improvements'. The causal model they offer looks like this:

4 I oversimplify and also will stretch the evidence base beyond what was actually available in order to highlight my point.
5 Or so much of the post evaluation maintains.

- The right people (those making decisions regarding under-nourished children) are targeted with nutritional messages.
- These people participate in project activities, and so are exposed to these messages.
- Exposure leads to acquisition of the desired knowledge.
- Acquisition of the knowledge leads to its adoption (i.e. a change in practice).
- The new practices make a substantial impact on nutritional outcomes. (White and Massett 2007, 630–631)

The HD evidence came from Tamil Nadu where INP was introduced in 173 out of 373 rural blocks in 11 districts,[6] and the evidence was positive. The World Bank reports it as one of the "most successful [projects] in the world in reducing malnutrition".[7] Let us imagine for the sake of argument that it was trialled successfully in a number of other places as well. The issue I want to raise is one about warranting that we do or do not have *enough of the right kind of variation.* In this case post hoc we have a far better idea of the answer than was available at the time. The point I want to note about this case is how many claims, of different kinds, go in to warranting the claim that the variation existing in the evidence cited was insufficient for warrant in this case.[8]

What went wrong in Bangladesh? Short answer: *Men* and *Mothers-in-law,* as White reports:

> The program targeted the mothers of young children. But mothers are frequently not the decision makers, and rarely the sole decision makers, with respect to the health & nutrition of their children [...] women do not go to market in rural Bangladesh; it is men who do the shopping. And for women in joint households—meaning they live with their mother-in-law—as a sizeable minority do, then the mother-in-law heads the women's domain. (White 2009, 6)

So in order to be appropriately varied, a warranting evidence base W(h) should have included trials in places where the mothers live in households where they have little control over what is bought and little control over who gets to eat what.

This creates a meta-problem. What would warrant this demand on variation? What does it take to defend that

> We have reason to think the evidence base needs this particular kind of variation?

6 https://extranet.who.int/nutrition/gina/en/node/23345 (last accessed: 17.9.2022).

7 https://documents1.worldbank.org/curated/en/939391468206956357/pdf/32563.pdf (last accessed: 1.7.2024).

8 Though again, we might have thought that there was sufficient variation and thought we had reasons for thinking that—significant ones of which turn out to be false.

Here are some of the kinds of evidence that went into arguing post hoc for the variation needed for the INP hypothesis to be warranted. Notice how detailed and varied they are.

- Data from the Demographic and Health Survey (DHS), which "show that only one in five women [in the BINP population] are solely responsible for decisions regarding their own health and that of their children, falling to only 1 in 10 of women living with their mother-in-law" (White and Massett 2007, 633).
- Conversations and reports from focus group meetings
  - "[...] from these came the memorable quote, 'we'll start doing that once our mothers-in-law have passed'" (White 2009, 6).
- Semi-structured interviews with 150 individuals (White and Massett 2007, 631)
- Ethnographical claims about gender in Bangladesh
  - E. g., Howard White reports that Sarah White's book, *Arguing with the Crocodile. Gender and Clash in Bangladesh*, played a significant role (White 1992).

This illustrates how much it takes to warrant that a set E(h) of evidence claims $\{e_1, \ldots e_n\}$ satisfies just one of the many requirements (e. g. the five listed above) to constitute a warranted evidence base W(h) for a hypothesis h.

Now couple this with the general need for auxiliary assumptions to join with h to imply e's of interest. Note that auxiliaries will generally be required for each of the e_i in W(h). So we need

$$h \ \& \ A_i \rightarrow e_i, \text{ where each } A_i \text{ is generally a big set } \{A_{i1}\ A_{i2} \ldots A_{in}\}.$$

Now think what this involves if each of the requirements above is to be satisfied. Warrant for h via HDB demands that $\exists \{\langle A_1, e_1 \rangle, \langle A_2, e_2 \rangle, \ldots, \langle A_n, e_n \rangle\}$ where

1. All A_{ij} and e_i are true.
2. n is large.
3. Pairs are sufficiently varied.
4. Enough pairs are novel.
5. Many are differentiating.

AND

6. All the above are warranted.

This is just the same kind of exploding network that we saw was required by the argument theory of evidence. And there is no place for any of the innumerable strands to end. *Plumping is ubiquitous.*

What about warrant? Just as with the argument theory, HDB allows us only conditional warrant. *Given* a set $\{\langle A_1, e_1 \rangle, \langle A_2, e_2 \rangle, \ldots, \langle A_n, E_n \rangle\}$ that satisfies conditions 1. to 5. with respect to h, then h is warranted. We are again left to 'decide' and 'commit' to such sets if we are to have any kind of warrant at all.

There are though theories of evidence that seem to avoid the problems I have raised, ones that take relevance to be a purely formal matter depending on no substantive facts. I think the only serious contenders like this right now are various Bayesian and other formal accounts that rely on probability relations to define relevance, generally involving facts about Prob(h/e) or Prob(e/h). John Norton in his work on the material theory of induction that I mentioned above provides a good analysis of the problems with these formal theories.

But even bracketing these drawbacks, these accounts do not get us out of the responsibility of choice. A probabilistic assessment of relevance requires minimally an event space and a probability measure over it. These both may be either objective or subjective. If either is objective, then we need warrant for our choices of them. Why, for example, is the particular event space $\{a_1, ..., a_n\}$ the right space relative to h—the space of events that bear on the truth of h? Warrant for h requires evidence that these are the right choices and evidence that what we offer as evidence for these is evidence. Then we have the same expanding net as in the two HD-based theories. On the other hand, either or both may be subjective. But this does not avoid multitudinous plumping. It just does it all in one fell swoop at the beginning.

5 Conclusion

The burden of voluntarism is great. Van Fraassen tells us: "[E]pistemic judgments are most like expressions of intention [...] [What] is conveyed [is] that I have made a decision, have formed an intention, am committed to a certain stance or program or course of action" (van Fraassen 2008, 254). Epistemologists sometimes suppose that we have epistemic obligations, presumably some appropriately nuanced version of 'Calibrate your beliefs to their warrant', and voluntarism will have implications for this.

In this conclusion, though, I want to direct our attention to what the implications of voluntarism may be well beyond epistemic obligations, to real moral, political and social obligations, on how epistemic voluntarism bears on our being 'committed to a certain stance or program or course of action', on what we should do in settings where our actions are subject to the constraints of morality. In such settings our actions should be conditioned by the facts, and part of the justification for those actions will be that we have tried to assure that our understanding of the facts is warranted. But if my arguments are correct, we never have but conditional warrant. Plumping is ubiquitous, and in adopting claims to guide our actions we could never know more than the tiniest iota of what we have plumped for. Nevertheless, we are responsible for our actions.

This can suggest great caution in our actions: "Even given the vast efforts of a huge epistemic community to constrain what can be plumped for at the cost of inconsistency, we have little reason for confidence in our understanding of the facts. So we should do little." But we know that often doing little or nothing is not the right course of action, despite uncertainty about what the real facts are or what our actions will produce.

We are in a situation analogous to one Eleonora Montuschi describes with respect to objectivity, that is elaborated in Cartwright et al. (2022)—*The Tangle of Science.* Or of those with a 'duty of care'. You are enjoined to *be objective* or you have taken a position that brings with it a duty of care. Some philosophers (Hacking 2015) have argued that 'objectivity' is an unusefully vague notion. Montuschi calls this familiar vague concept, *objectivity as we know it* (OWKI). OWKI does not admit of one clear definition (even a disjunctive one) that is appropriate in most contexts where we want the term to do work. So, Ian Hacking and others argue, it should be replaced context by context with a tighter concept that makes the precise demand we intend in that context.

Montuschi objects to wholesale replacement. As argued in *The Tangle*, that's because often we want this familiar concept—OWKI—precisely because it is vague and, as Bernard Williams put it, like many vague everyday concepts (e.g. 'poverty' or 'greedy') it is both 'world-guided' and 'action guiding' (Williams 1986). In many contexts we use OWKI to assign responsibility for *objectivity to be found* (OTBF). We do not know before-hand what will—or should—matter in the context. In enjoining an agent to 'be objective', we place on that agent the duty to *figure out* what objectivity consists in in that setting and to instantiate it.

This is like taking a job—like financial adviser to customers in a bank or health-care practitioner—that brings with it a legal duty of care. Sometimes what this consists in is spelled out in detail. But very often it is intentionally left vague because no such list will cover all the possibilities. The phrase has real bite—it is used by courts and by professional disciplinary bodies as the basis for deciding matters of liability and misconduct. It appears in legislation. But what counts as the proper fulfilment of that duty depends on the context, e.g. the person to whom the duty is owed, and it is part of the duty that the agent endeavour to figure out what that duty consists in case by case.

I propose that the conclusions about conditional warrant put us in the same situation with respect to our fact-guided actions as are those given the vague injunction to be objective or who are landed with a duty of care. Once we have plumped, there are rules for determining what fact claims are warranted. But as with 'objectivity to be found' and many instances of duty of care, there are no rules for how to plump. Yet in situations where our actions are subject to the constraints of morality, we must do this responsibly and as well as possible, all things considered, and we are rightly subject to blame, post hoc, for certain kinds of mistakes.

For instance Barrotta and Montuschi (2018) tell of the Vajont dam disaster, where an entire town in the Dolomites was wiped away by a gigantic wave of water because a massive limestone landslide fell into the reservoir, the dam resisted the impact and the overflowing water flooded the entire valley. The engineers designing the dam expected that no such landslide would occur. In doing so they relied on a well-supported general principle to tell them about local rock: From the chief engineer, "[...] the rocks [of the Veneto region] are generally very good [...]. Overall, limestone is honest because it reveals its flaws on its surface" (Barrotta and Montuschi 2018, 25). In-depth geological studies were considered unnecessary because the rocks of the area did not raise visible concern. Tragically, there was evidence available at the time that this principle was not true of the area rocks. There was clear local knowledge of large limestone landslides up the valley. Barrotta and Montuschi fault the engineers for neglecting this local knowledge and for not doing a geological study given the scale of the disaster that could (and in this case did) result from, in van Fraassen's words, 'committing' to the general claim abut limestone that might prove false in a particular case. The engineers were morally wrong, all things considered, in taking the general principle to be locally reliable without more investigation given that lives were at stake. In my words, they plumped in the wrong way given the actions—their plans for how the dam was to be built—that ensued from the claims they plumped for.

How shall we, unguided by rules, make these difficult epistemic decisions in morally sensitive settings where they are action guiding, in intersection with a variety of other morally relevant considerations? That is an on-going problem in philosophy as in life, about which even those philosophers who advocate morality without rules[9] have not had enough to say. Still, we often manage, and moreover, we must. As Amartya Sen urges: "The problem here isn't fearing that scientific action *might be* value-loaded, but fearing that it *might not*. Value-loading here is not so much a right as a duty" (Sen and Lloyd 1983, 104).

9 Like the so-called 'Oxford pessimists' (e.g. Isaiah Berlin, Stuart Hampshire, Bernard Williams), those who advocate moral particularism or those who base morality on the moral sentiments but do not go down the route of codifying procedures (like the veil of ignorance).

References

Austen, J. 2006 [1811]. *Sense and Sensibility*, ed. E. Copeland. Cambridge: Cambridge University Press.

Barrotta, P. and Montuschi, E. 2018. The Dam Project: Who Are the Experts? *Science and Democracy: Controversies and Conflicts*, ed. P. Barrotta and G. Scarafile, 17–34. Amsterdam: John Benjamins.

Cartwright, N., Hardie, J., Montuschi, E., Soleiman, M. and Thresher, A. C. 2022. *The Tangle of Science: Reliability Beyond Method, Rigour, and Objectivity.* Oxford: Oxford University Press.

Di Bucchianico, M. E. 2009. *Modelling High Temperature Superconductivity: A Philosophical Inquiry in Theory, Experiment and Dissent.* Doctoral dissertation, London School of Economics and Political Science.

Hacking, I. 2015. Let's Not Talk about Objectivity. In: *Objectivity in Science*, eds. J. Y. Tsou, A. Richardson and F. Padovani, 19–33. Dordrecht: Springer.

Hempel, C. G. 1965. Aspects of Scientific Explanation. New York: The Free Press.

Neurath, O. 1983. Protocol Statements. In: *Philosophical Papers 1913–1946*, 91–99. Dordrecht: Springer.

Norton, J. D. 2014. A Material Defense of Inductive Inference. Manuscript. https://sites.pitt.edu/~jdnorton/papers/Skepticism_Induction.pdf (last accessed: 7.1.2024).

Reiss, J. 2015. A Pragmatist Theory of Evidence. *Philosophy of Science* 82(3): 341–362.

Scriven, M. 1966. *Applied Logic: An Introduction to Scientific Reasoning.* Palo Alto, CA: Behavioral Research Laboratories.

Scriven, M. 1976. *Reasoning.* New York: McGraw-Hill College.

Scriven, M. 1981. *The Logic of Evaluation.* Point Reyes, CA: Edgepress.

Scriven, M. 1998. Minimalist Theory: The Least Theory That Practice Requires. *The American Journal of Evaluation* 19(1): 57–70.

Sen, A. 1983. Accounts, Actions and Values: Objectivity of Social Science. In: *Social Theory and Political Practice*, ed. C. Lloyd, 87–107. Oxford: Clarendon Press.

van Fraassen, B. 2008. Belief and the Will. In: *Reasoning: Studies of Human Inference and Its Foundations*, eds. J. Adler and L. Rips, 47–60. Cambridge: Cambridge University Press.

White, H. 2009. Theory-Based Impact Evaluation: Principles and Practice. *3ie Working Papers.* https://www.3ieimpact.org/evidence-hub/publications/working-papers/theory-based-impact-evaluation-principles-and-practice (last accessed: 24.11.2023).

White, H. and Masset, E. 2007. Assessing Interventions to Improve Child Nutrition: A Theory-Based Impact Evaluation of the Bangladesh Integrated Nutrition Project. *Journal of International Development: The Journal of the Development Studies Association* 19(5): 627–652.

White, S. C. 1992. *Arguing with the Crocodile: Gender and Class in Bangladesh.* London: Zed Books.

Williams, B. 1986. *Ethics and the Limits of Philosophy.* London: Routledge.

Bas C. van Fraassen

Reply to the Conference Participants

Abstract: After a brief note about Henri Lauener and Karel Lambert, these replies engage the participants' illuminating arguments and views.

The Lauener Prize for an Outstanding Oeuvre in Analytical Philosophy is a great honour for me. I am thankful to the Lauener Foundation, and especially to Professor Dagfinn Føllesdal, to Dr. Michael Frauchiger, and to the other members of the Lauener Foundation Council. I feel all the more honoured when I reflect on the distinguished array of previous philosophers awarded this prize. They all shine in my philosophical firmament. One of my heroes in philosophy was Hilary Putnam with whom I had a number of debates, starting at the first meeting of the PSA in 1968. They include also Tim Scanlon, a former colleague at Princeton, and others whom I knew and admired greatly, such as Michael Dummett and Ruth Marcus.

While I never had the opportunity to meet Henri Lauener, I do actually have a personal connection to him, at second remove. For Professor Lauener was greatly admired by my logic teacher Karel Lambert, with whom he spent a good deal of time not only in Switzerland but also in California (see Lambert 2002). Shortly before the conference I spoke about this with Lambert, who is now retired, and who was delighted to recall their conversations. Being philosophers, those conversations were argumentative, small debates in themselves. I can only try to imagine what they could have been like, between one very pragmatic American philosopher and one Continental philosopher who developed a transcendentalist philosophy.

Although Lambert was my beloved first philosophy teacher, I think I would mostly have sided with Lauener. Lambert described Lauener as decidedly an analytic philosopher, with a great love both for Quine and for Quine's adversary Ruth Marcus. But Lauener's philosophy had other features as well, with which I would have been more in sympathy than Lambert was. I mean Lauener's "Kantian humility" about our limited cognitive reach, his critique of philosophical naturalism, and the central role he gave to norms. In philosophy of science too I can point to his acceptance, which I count as entirely right, that all observation language is theory-laden.

Participation in the conference organized together with the award ceremony was a genuine intellectual pleasure for me, because of the constant challenge it

https://doi.org/10.1515/9783111019802-014

provided. I can hardly do justice to its spirit of free discussion, or adequately express my thanks, with the short replies I can give here.

Reply to Steven French, "Representation and Realism: On Being a Structuralist All the Way (Up and) Down"

There might at first blush seem to be a radical disconnect between Steven French's philosophical orientation and my own, with mine unblushingly empiricist and his unblushingly metaphysical. But as French's essay shows, and as is surely clear from much of our past interaction, that is simply not so. While there is a salient difference between us within meta-philosophy, in philosophy of science we encounter by and large the same problems. And we do so in a way that allows for his reasoned quarrels with empiricist views and mine with ontic structuralism, when it comes to the interpretation of such mystery-engendering intellectual constructs as quantum theory.

Here I will focus on two issues. The first concerns how theory relates to experience, and the second concerns criteria of completeness putatively normative in scientific practice.

That first issue French illustrates with a quotation from Hans Halvorson, attributing to the semantic approach the contention that "a theory is a class of models". What Halvorson criticized (and as French emphasizes, had been rejected explicitly well before) is the letter of that identification, while Halvorson himself stayed adherent to its spirit. That we could sum up with Kant's contention in the Preface to his *Metaphysical Foundations of Natural Science:*

> [I]n any special doctrine of nature [...] there can be only as much *proper* science as there is *mathematics* therein. (Kant 2004 [1786], 6)[1]

Against this (and more in line with other Kantian insights), although a scientific representation may be a mathematical structure, it does not "follow that the identity of a theory can be defined in terms of the corresponding set of mathematical structures without reference to their representational function" (van Fraassen 2014, 278). French agrees, but then immediately recognizes the corollary question: "Granted that [...] there is more to a theory than the mathematical structure [, that]

1 "Ich behaupte aber, daß in jeder besonderen Naturlehre nur so viel *eigentliche* Wissenschaft angetroffen werden könne, als darin *Mathematik* anzutreffen ist" (Kant 1977, 13).

immediately raises the question: how are we to capture this representational function […]?" (88)

Although the question "what is representation?" cannot be answered solely or simply in terms of resemblances of any kind (whether isomorphism, partial isomorphism, homomorphism, or …), it is certainly the case that many forms of representation, specifically in the sciences, *trade on resemblance* for their success. The relation between data models, which contain the empirical information about the pertinent phenomena in summary terms, and theoretical models offered as representation of those phenomena, surely demands some such relationship. And these sorts of relations are just what go into the paradigm examples of what constitutes structure. But to study these is to remain just within the *mathematical form* of the theory, its models, and its data structures. That does not address the question of how we are to capture that representational function.

We are agreed that something more is needed, and French and I have expressed parallel worries. For French the question is whether this something is something non-structural, at the risk of losing the pristine purity of radical ontic structuralism itself. For me the question is instead whether empiricist structuralism, the view that scientific representation is a representation of phenomena as embeddable in certain abstract structures, describable or identifiably codifiable only up to isomorphism, can be maintained as all by itself a view of science. To start, we are also content with roughly parallel answers, to the effect that acknowledging the indexical element in practical knowledge, the act of self-location in a theory's logical space to make that theory applicable to our own situation, alleviates these worries.

Nevertheless, such an answer has its challenges. I want to briefly address two. The first takes us eventually well outside philosophy of science; the second is the "Appearance from Reality" problem.

First then, French raises the possibility of an objection, namely that introducing self-location would (or does) add a non-structuralist element to his ambitious ontic structuralism. The same putative objection can be raised for empiricist structuralism. We agree that the objection is met by a careful reflection on just what is added. Self-location, though ineliminable, does not introduce the ineffable, nor postulate phenomena of consciousness that would of logical necessity remain outside the domain of any science.[2]

Acts of self-location, such as occur every day, can certainly be part of the subject matter of a scientific theory. They can, but we have to add two points that will induce a residual tension. The first is that the pertinent scientific models of such phenomena would also be purely structural. And the second is that those models

2 See my discussion at an earlier juncture in the realism debates, in Section 1 of van Fraassen 1993.

would count as representations only in the sense that they can or could be used to represent those practical-intellectual phenomena. Taken together these may sound a bit paradoxical, as if we are taking away with one hand when giving with the other. But it does not by itself yield an inconsistency.

Nevertheless, it is not enough to appreciate that. That response has the right form to be applicable in any *particular* case, such as our self-location in a model of climate change in order to generate pertinent predictions for ourselves. But what is implied about the resources we have for doing that, in any such case whatsoever? These resources require our having a language within which we experience, in principle, no limits of expression. Any attempt to form a representation of *that*, whatever provides us those resources taken altogether, is beset by all the logical and semantic paradoxes that came to light since the days of Cantor and Frege. The great shock we experienced in metatheory since those days is that there are limits we can see *from within*, precisely in the sense that we realize that we cannot adequately describe them *from without*—in that sense we, who incarnate our language, are beyond the very possibility of representation (cf. van Fraassen 2019).

I do not argue this to object to French's, or my own, alleviation of the above worry, but to insist that we cannot misunderstand local solutions as reaching further than they do. In this area, however, we are no longer in philosophy of science, we are where logic and metaphysics become increasingly difficult to distinguish, and certainly where empiricist anxieties become acute.[3]

The challenge to which French devotes more of his discussion concerns the second, Appearance from Reality (putative) criterion: "that the scientific representation of nature should include the appearances by showing how they are 'produced as a proper part of the depicted reality'" (93). The difference in our philosophical orientation comes to the fore here. For an empiricist the question of whether this criterion is to be met in scientific theorizing is a question about the norms, or ideals that have a normative function, for theorizing, that hold in scientific practice. That the criterion is to be met in fact would seem to be axiomatic for ontic structural realism. There it is not a matter of norms in practice but of what is required in developing an ontology. If structuralism is to hold all the way up and down, the appearances must derive from the structure that constitutes reality; the question is not whether they do, but how.

3 Recall Kant, on arguments "by means of which we can infer from something with which we are acquainted to something of which we have no concept, and yet to which we nevertheless, by an unavoidable illusion, give objective reality" (*The Critique of Pure Reason*, Kant 1998 [1781/1787], A339/B397, 409). Kant's critique had its sequel in the foundations of mathematics, not just positively in Brouwer's demonstration ("the ontoereikendheid van the taal") but negatively in the many unsatisfactory attempts to escape semantic paradoxes within "classical" metatheory.

The "how" needs to be a structural version of production, in the relevant sense. As I understand it, appearances are the contents of measurement outcomes. There are good examples that fit what is needed for ontic structuralism in this respect. The example I gave in the book, of Copernicus' derivation of the appearances of planetary motion from the postulated reality in his model, seems to me a prime example of a successful structuralist derivation of the appearances. A structure is derived from a structure, in a kinematic version of projective geometry.

Another example that seems to me to be assimilable to a structuralist account is the explanation of the photoelectric effect. This phenomenon was studied experimentally in the nineteenth century, and by the end seemed to be explicable as due to a transfer of energy from light waves to electrons. As expected, the number of emitted electrons (derived from measurements of the resulting current) increases with the intensity of the incident light. But the implied relation between the intensity of the light and kinetic energy of the emitted electrons was not borne out in measurement. The new explanation, now familiar to us, was due to Einstein taking up Planck's suggestion about discrete transfer of energy. The account clearly takes more than a geometric kinematics, but can still, I think, be told as a story about the structure of electromagnetic fields and energy transfer.

French's examples of the role of the Exclusion Principle and Permutation Symmetry are also apt and pertinent. But to maintain ontic structural realism more is needed than great examples that fit the mould, however well. So French is right to explore general ideas about counterfactual dependence and non-causal dependencies, which do not feel alien to his project for ontology. Is it possible to find there a general characterization, in structural terms, of the production or derivation of the appearances from the postulated reality?

What I take to be the main obstacle is the theory of measurement in quantum mechanics, as French points out. From the traditional point of view, the theory presents a paradox, precisely because the appearances (that is, the contents of actual, specific measurement outcomes) do not derive in any relevant way from the reality (that is, the quantum-mechanical states and their dynamics). At least, they do not do so in standard quantum mechanics. They may in an alternative theory, and some such theory may supersede quantum theory, but that is still a "may be". French takes up several possibilities here but wisely, to my mind, does not endorse any. His conclusion, at that point in the discussion, bluntly states the dilemma:

> either we exclude such explanatory assumptions from our specification of "the" theory, and then bite the empiricist bullet when it comes to obtaining the appearances from reality, or we introduce some such assumption and accept both the local and global costs. (101)

What is more to the point, as far as our philosophical differences go, is how different our take is on the Appearance from Reality problem. French addresses the question of how to develop an ontology, ontic structural realism. So his concern is with the question for ontology: what is the structure of the universe, and here the sub-question, of how the appearances belong to what is real, cannot be avoided.

Within an empiricist view of science, the concern is very different, the question is not about what the world is like but about what science is like. The question is whether the completeness ideal, in which all appearances are derivable from the postulated reality in the Final Theory, plays an inescapable normative role in scientific practice. I submit that the answer is negative, as shown in the actual history of physics, since the Copenhagen physicists had no hesitation in developing a theory that fails to meet that criterion. Moreover, the theory they developed in which individual measurement outcomes remain spontaneous, was happily embraced in practice by a large majority of working physicists as well. Practice belied that traditional imperative, to aim for the sort of completeness the Appearance from Reality criterion calls for.

However, there is a surprising, and to me welcome, turn at this point in French's deliberation. Having canvassed those interpretative extensions of quantum theory that accept "the local and global costs" (101), French explores the possibility of a different approach to theoretical physical science, one that could sidestep the dilemma.

This is a subtle and delicate move, in which we step outside the level of semantics and ontology, into a broadly pragmatic—or perhaps more accurately, phenomenological—approach to what is meant by our talk about scientific theories. Talk of theories gives way to talk of theorizing as scientific activity, taken to encompass reflective practices of the philosophers as well as constructions by physicists. This is intriguing, tantalizing, signalling perhaps a sea change in French's philosophy. I look forward to learning much more about this, which may bring views that mark a new stage for both of us, beyond what the realism-empiricism debates have been so far.

Reply to Michela Massimi, "The Perspectival Nature of Scientific Representation"

The perspectival realism project is very rich, as it brings to light many and diverse problems and issues for realists in philosophy of science. What it promises is a new scientific realism that is thoroughly in tune with scientific practice and incorporates insights about representation gained in the study of the visual arts as well as of modelling.

Obviously I cannot join the project, since it is a project for scientific realists. But I have to ask myself: do these issues and problems for our understanding of the sciences not arise, mutatis mutandis, for an empiricist view of science? To what extent is an appreciation of perspectivity in representation already present in an empiricist view, and to what extent must it figure in the improvement of such a view?

In my own thinking about scientific representation so far I have highlighted perspectivity in just two places. First of all, the appearances, by which I mean the contents of measurement outcomes, are perspectival. Secondly, the defining difference between pure and applied science is that an application of scientific theory for prediction, control, or manipulation involves an act of self-location.

The first was emphasized by Poincaré: a measurement outcome does not show what the target is like, simpliciter, it shows what the target is like "from here", that is, within the measurement set-up. As a guiding example of a measurement outcome we could take the fighter pilot's shout "Bogies at 2 o'clock high!" It is then possible, of course, to construct a model of the measurement that comprises both set-up and outcome. In that model the perspectivity disappears in favour of relationality. But the point stands that the appearances are perspectival.

For the second way my guiding example is Kant's image of someone consulting a map of the heavens to predict where the sun will rise. To do that one must locate oneself with respect to the map, identifying both position and direction. The point generalizes to the use of models. To make a practical prediction—carry an umbrella or not?—on the basis of the NOAA Graphical Weather Map for North America a person needs to self-locate on that map. But again we have to distinguish carefully where the perspectivity comes in. There is no self-location supplied together with, or given in, the map. That map is a graphic representation of a model of the atmospheric conditions in that area, in which no point or direction has a privileged role. The self-location required is an act by the user required for practical application.

So far then, these are the only ingressions of perspectivity in the discussion of models and theories. Is that enough? Or will an empiricist view find itself as sorely pressed, as scientific realisms clearly are, when we wind our way through the thicket of issues in the literature on perspectival realism?

To begin let me focus on this idea, which would be at odds with the limited role of perspectivity that I outlined above:

> (*) Models, or theories, themselves are perspectival, in that they present their target in a certain way, that makes it apt to say that the representation is from, or relative to, a vantage point, or point of view, or frame of reference in some generalized sense.

This thesis (*) is what Michela Massimi calls *the perspectivalist assumption*, understood in terms of *perspectival$_1$* (a representation "from a particular point of view") rather than perspectival$_2$ representations.

On the face of it, as Massimi points out, the perspectivalist assumption is in tension with the representationalist assumption, central to scientific realism. As she presents it, the tension becomes acute if different perspectives are understood as attributing mutually incompatible properties to a common target. Thus understood Massimi rejects the perspectivalist assumption, for she is developing a view that can be aptly called a perspectival realism. In response she is developing an alternative conception that trades on the perspectival$_2$ representation.

However, this leaning toward scientific realism I do not share. The representationalist assumption is clearly not part of any empiricist view. So we should ask: might (*) be acceptable when realist assumptions are not in play?

To be charitable, we should understand this question as a challenge to find ways to understand (*) in some fruitful and acceptable way. "Perspective" and "perspectival" have metaphorical uses related to visual perspective, and since metaphors are infinitely extendible, there might be various ways to do this.

(*A*) In (*) the terms "vantage point", "point of view", "frame of reference" are also used metaphorically or analogically. Sometimes in such a context I see "relative to a theory" (a particle model of a fluid is a good model relative to statistical mechanics). Is that something we could insert in (*) to replace those terms? It does not seem so. It would not do to say that theories present their targets in a way that is relative to a theory, for that would be circular or vacuous. On the other hand, to say that a model presents its target relative to a theory could only mean that it is a model of that theory—so, also not helpful.

(*B*) Perhaps we can explore the fact that different theories may represent the same phenomena in different ways, as is illustrated by many examples. Phenomenological thermodynamics and statistical mechanics provide different models of fluid and gas behaviour, and represent the quantities that characterize them differently. For a realist this seems to present a problem, at least at first blush, and certainly does—as Massimi argues—if different perspectives are understood as attributing mutually incompatible properties to a common target. But from an empiricist point of view that does not present a problem, for the empirical conditions of accuracy refer only to the measurable aspects of those phenomena.

(*C*) Scientists immersed in the different theories see the phenomena through different theory-coloured glasses. In that sense we can say, with an easily understandable metaphor, that they see the phenomena from or in a different perspective. That is a helpful point about the phenomenology of scientific practice, famously explored, in different ways, by Husserl and Kuhn. But it does not seem to me that the metaphor plays an indispensable role.

(*D*) That the same theory admits generally of different interpretations, that is also right. Quantum mechanics and classical electromagnetic theory present us with different representations of the interaction between light and matter, but in addition, the quantum-mechanical representation can itself be understood in different ways. These ways are provided for us by, for example, Everett, GRW, Bohm, and so on. But these "interpretations" are themselves theories, ones that extend the official quantum mechanics by adding empirically superfluous variables, or modifying it in ways that make no difference to how the theory fits the data. That such alternative interpretations can be feasible is an important point for our understanding of the sciences. Do they motivate some recourse to the "perspective" metaphor? Perhaps they do for a realist who holds that there must be a true account, even if we do not know what it is, that matches one of the feasible interpretations. But this point about indeterminacy on the level of theory-interpretation holds for theories generally, without reference to truth-values. Perhaps, in analogy to the previous point, we could say that adherents to those different interpretations—the Everettians, the Bohmians, etc.—see the phenomena through not just theory-coloured but metaphysics-tinted glasses.

None of these give us any reason to assert (*), or, therefore, to endorse the perspectivalist assumption, understood in terms of the perspectival_1 relation, at least not in any non-trivial sense. This conclusion, on the side of an empiricist view, seems to me to strengthen Massimi's reasons to pursue the different perspectival_2 relation instead.

The introduction of perspectival_2 representation as a clue to the character of scientific representation in general is thoroughly intriguing. In the present article Massimi does not go on to develop this in detail, though we may note that she has made it clear elsewhere that it builds on Kant's notion of the *focus imaginarius* (Massimi 2017). If Massimi shows that the representationalist assumption is compatible with the perspectivalist assumption, newly understood in terms of the perspectival_2 relation, then the next question for me will be whether the latter can have a counterpart in an empiricist view of science. Intriguing, tantalizing, ... I am looking forward to this.

Reply to Finnur Dellsén, "Expanding the Empirical Realm: Constructive Empiricism and Augmented Observation"

Finnur Dellsén presents an objection to constructive empiricism based on the possibility of epistemic augmentation (by immigration or upgrading) in which the range of observability is extended. That is a very interesting possibility, which deserves careful examination (see also van Fraassen 2005). Although it is not a possibility that we are currently confronting, it is important to see whether an empiricist view of science might need to be amended, or would admit feasible alternatives that are also empiricist rather than scientific realist (see also Dellsén 2017). The empiricist tradition can only benefit from a construction of alternative answers to what empiricism can be now.

This will require understanding what a case of epistemic augmentation would be like. But let us begin with Dellsén's objection to the view that science aims to give us empirically adequate theories.

The Main Points, and Their Import

First of all we must clearly concede a great deal to Dellsén:
- the limits of observability depend on what is possible *for us*, and would therefore be extended if our epistemic community changed through epistemic augmentation,
- the criterion of empirical adequacy is defined in terms of observability, and would therefore become stricter if our epistemic community changed through epistemic augmentation,
- if a theory was empirically adequate before such a change, it might no longer be empirically adequate after the change, and equally, any theories constructed after such a change would have to meet a stricter criterion of empirical adequacy than they would have had to meet before that change,
- so, it would be harder and less likely for us to have empirically adequate theories after epistemic augmentation.[4]

The question is now whether this presents a problem for constructive empiricism, which was defined as follows:

> (*A) *Science aims to give us theories which are empirically adequate; and*
> (*B) *acceptance of a theory involves as belief only that it is empirically adequate.* (van Fraassen 1980, 12, with the marked division inserted)

Put bluntly, do Dellsén's points mean that scientists would, according to (*A), be doing the scientifically right thing if they rejected or prevented epistemic augmen-

4 We could quibble with this last step, for the effect of the new challenges might be fecund new theoretical activity likely to bear fruit.

tation, in order to make it more likely for their scientific activity to give us theories which are empirically adequate?

That inference would require an extra premise:

> If science aims to give us theories which are empirically adequate, then doing anything at all that would make it more likely that we will have empirically adequate theories will be doing the scientifically right thing.

But with this extra premise in place we would also have much that seems unacceptable. For it implies that scientists would be doing the right thing if they stuck to exceedingly prudent, un-audacious, even vacuous or tautological theories. And a parallel objection could be launched against the scientific realist view that science aims to give us true theories, for that kind of prudence would also make it more likely for their scientific activity to give us true theories. These consequences seem to me to make that extra premise implausible, reflecting an implausible reading of (*A).

What is at fault here? It seems that (*A) is given an exclusionary reading, which is not the default reading for this form of statement. An exclusionary reading would not be a plausible construal of "*X aims to give us Ys which are F*" in other contexts. Such an assertion could not in general be reasonably construed to involve a denial that *X* was aiming to give us *Ys* which are also satisfactory in other respects.[5]

Notice that the second part, (*B) could also be given an exclusionary reading. But the paragraph following it in the text spells out explicitly that it was not meant to be read in that way. The assertion "acceptance of a scientific theory involves as belief only that it is empirically adequate" is followed by pointing out that this does not *equate* acceptance with that belief, it does not deny that something else is involved as well. Acceptance has a pragmatic dimension, it involves more, it involves commitment, it involves immersion in the theory, a way of approaching phenomena in the terms of that theory. Similarly, a non-exclusionary reading of (*A) will leave room for science to be aiming to give us empirically adequate theories that are also satisfactory in other respects. Specifically, there are pragmatic—non-epistemic—virtues as well, such as informativeness, simplicity, beauty, explanatoriness.

But in that case, what is the role played by the specific mention of empirical adequacy in these statements, on a non-exclusionary reading? First of all, it introduces the important contrast with the scientific realist criterion of success: science

5 The possibility of unintended readings of one's text may remind one of the famous horror story "The Monkey's Paw".

aims to give us empirically adequate theories *rather than* true theories, and acceptance involves belief in empirical adequacy *rather than* belief in truth. Secondly, its prominent mention indicates that here we are speaking of the bottom-line, as businessmen would say, the virtue whose lack makes the others lose their importance.

None of this implies that doing just anything to make it more likely that we have theories which have that virtue is the scientifically right thing to do. Thus, I submit that logically, the mooted bad consequences for the constructive empiricist view of science, do not follow.[6]

However, a defence based solely on a logical point of order is never convincing by itself. And Dellsén is entirely right that we should take the possibility of epistemic augmentation seriously, and to insist that we explore what it means for an empiricist understanding of the scientific enterprise. So let us do that.

Scenarios for Epistemic Augmentation

My plan here is to sketch a scenario of epistemic augmentation. It should be of a sort that could be hospitable to the objection Dellsén presented, a scenario in which something that was empirically adequate no longer is so.

The first ingredient is the imagined community C that is the communal narrator, that plays the initial role of *us*. Let it be one in which everyone is agnostic about the unobservable and accepted theories are only believed to be empirically adequate. Now a new community C# appears.

There are too many unsettling questions to be asked about how we could come to accept dolphins or extra-terrestrial aliens as persons. So let's say at once that in C the members of C# are accepted as persons, and taken to be humans, humans but not like us. As guiding example take the children in Arthur C. Clarke's *Childhood End:* they are human since they were born of humans, but they are not like us, and their behaviour appears to show that they can access ways things are differently from us, while a good deal of communication between us and them appears to work well. I do not mean to import any further details from that novel, only this.

Now suppose the C# announce: "We can observe Z", and Z is nothing that counts as observable for us in C. To say even this much carries some presuppositions. Do we in community C understand this announcement? For that we need to have the relevant concept of Z. It has to be a theoretical concept if we count Z as

6 Related problems were discussed by Cartwright (2007) and Percival (2007), and my response to them on pages 338–344 in Monton (2007).

not observable. To make it a serious matter for us, it must be a concept in one of our accepted theories, call it *T*.

For now let us add the assumption—but keep in mind that this may be too strong—that we have an instrument *I* that is classified relative to *T* as measuring or detecting Z.[7]

Secondly, given the puzzles the C# behaviour already presented for us, how should we understand "observe" in this announcement? No use just now to try to solve problems about consciousness in philosophy of mind—let's at least insist on some minimal common ground. Can we at least believe that they, bodily, materially, can function as a measuring instrument for Z, relative to *T?* This we can check empirically, correlating their reports with measurements with our instrument *I*. Being able to do this we can at least support the contention that what they say *means what we understand* by the same statement, and more precisely, the contention that their reports of observing Z are genuine observation reports.

You see, we had to build some important assumptions into this story so far, to make it an intelligible one. We can actually reinforce its status, perhaps even plausibility, because we know of something a bit like that in our actual world. Pigeons have long been known to be excellent navigators, and the theory that is currently accepted is that they do so by sensing the earth's magnetic field. Experiments on pigeons, of the usual sadistic sort, have led to the conclusion that the pigeon's magnetic sense is located in the nasal region, and possible because of the presence of magnetite crystals there. So here we have a clue already to how we in C could begin to construct a theory to account for the correlation between our instrument I readings and the C# "observation" reports.

The next step, now at least not unreasonable, would be for us to believe not only that the C# "observation" reports are reliable indicators of something, but that they are genuine observation reports. Let's pass silently over any deeper problems that could be involved here, again to let sleeping philosophy of mind puzzles lie.

But now there are two ways we could do that. We could be literalistic and say: when the C# say they observe Z, then Z is present, and hence is real. Something which was unobservable for us, in community C, is observable now that we are members of the merged C-C# community.

I think that might be what is envisaged when various writers have offered examples of epistemic augmentation. But it would beg the question to be so literal-

7 It is important to keep in mind that theoretically postulated entities and quantities are not detectable or measurable directly; the relevant measurement operations involve calculations based on the theory itself—what George Smith calls theory-mediated measurement (cf. van Fraassen 2012). See further my response to Michael Friedman.

istic! To stay in character we should take seriously the theory-laden-ness of such reports (cf. e.g. van Fraassen 1993). For example, if a person raised and educated in the phlogiston theory were to shout "Lo, phlogiston escaping!", that would be a genuine observation report, and very reliable as an indication of the presence of fire.

In view of this we should say: "When the C# say they observe Z, that has exactly the same status as our report that our instrument *I* indicates the presence of Z."

Thus it is true that the C# observe something that we who only belong to C cannot observe (but only check by measurement), namely that we are in a situation correctly classified *relative to T* as including Z. For the use of *I* ending with the result that Z is present is a theory T-mediated measurement.

There is, in this understanding of the scenario, as yet no threat to the empirical adequacy of *T.*

But that could come next. For now the C# could make reports of a sort that we had not ever actually obtained in our sparse use of instrument *I* (but which we can now check with *I* and get "confirmation") and which violate *T.* For example, if Z was in fact the earth's magnetic field, certain data provided by the C# about the magnitude and direction of the magnetic field at a certain location might perhaps be, even with any reasonable auxiliary hypotheses, impossible to fit into Lorentz' force law. An anomaly, of the sort that refutes theories, by refuting their empirical adequacy!

Notice though that these results could have been gotten by use of instrument *I.* So it is not a telling case—although we did not realize it, in this case *T* was never empirically adequate to begin, though we believed it was.

So we are not finding any obstacles here to a merging of C with C# in which the C would maintain their agnosticism with respect to the theoretical entities and quantities in their accepted theories.

The main reason for this may be that I introduced the assumption that in C we have an instrument which, relative to our accepted theory *T,* counts as measuring Z. The epistemic augmentation envisaged consisted simply in adding the possibility of direct unmediated observing of a sort that correlates precisely with observations of an instrumental interaction. Not a radical enough epistemic augmentation to lead to any problematic consequences.

If we now go back and take instrument *I* out of the story, we lose the empirical data that could have led us toward believing that the C# reports should be treated as observation reports.[8] A new scenario without that would have to include an in-

8 We could replace the use of a single instrument *I* in this story with the investigation of a range of

ductive step without empirical support. I do not want to deny the possibility of rationally responsible leaps of faith or radical conversion. That may not be implausible as a story if in all non-scientific respects the members of C and C# seem to be enjoying perfect communication, develop some empathetic ways of understanding each other's way of being.

But though I have not thought through a carefully constructed scenario of that sort, I do not see any reason to think that in that case the C would not also be able to maintain their scientific agnosticism.

Reply to Milena Ivanova, "Aesthetic Virtues and Theory Acceptance"

Theoretical virtues present a delicate subject for philosophical discussion for, as Ivanova points out, scientists themselves have contributed an impressive body of evidence for the role these virtues play in practice. Elegance, unity, coherence, simplicity, and explanatory power all contribute to the prior plausibility and increasing confidence a theory will enjoy in the scientific community. And it is hard, a matter of dry logic rather than felt conviction, to drive a wedge between the scientists' judgement of what is right and the impartial question of what is true.

Milena Ivanova's critique of realists' take on the theoretical virtues is incisive. More: she takes the wind out of their sails by placing the issue in the context of the cognitive activity of the scientific actors involved. I can only offer a few footnotes to support her line of argument.

1) There would seem to be a tension between the argument that scientists are inspired by the desire for theoretical virtues such as beauty and elegance, which guide them in the choice of theories and research, and the argument based on how the track record shows that such virtuous theories have been successful. For if the former is correct then the actual history of science must be offering us a biased sample: the more virtuous theories were more likely to be pursued, developed, and more likely to elicit the devotion and perseverance of the scientists involved. But if the history's record is a biased sample in those respects, then a predominance of beautiful, elegant theories among those that were successful is not so telling. The ugly ones did not have the same chance of being developed so as to meet the empirical challenges.[9]

observable effects in the pertinent conditions, or a complex experimental set-up; that would not make a difference here.

9 I owe this argument to a critique of McAllister (1996) by Jill Sigman.

2) Salmon's argument is at first blush more substantial, but it also suffers from some internal tension. Salmon takes Kuhn's list of five virtues as instrumental in scientific practice. He points out first that they are of two kinds. Three of them, accuracy, scope, and fruitfulness, he claims to be understandable in terms of likelihood. But simplicity and "consistency" (in the sense of "fitting in well" with other accepted theories in the environment) are subject to judgement that cannot be reconstrued in terms of probability of truth. So, conceding this to Kuhn, Salmon holds that their role lies in shaping of the scientists' prior probabilities of competing hypotheses. These prior probabilities, plus the new evidence adduced, then determine the posterior (comparative) probabilities of those hypotheses, via Bayes' theorem.

The obvious question for Salmon's account is whether the scientists' higher prior probabilities, shaped by desires for simplicity and "consistency", reflect a higher probability of truth. Those desires are, by his own account, external to any factors that he can argue to make a hypothesis more likely to be true.

Salmon's response to this challenge is to align himself with Patrick Suppes'—surely correct—insistence that

> [w]e generally bring to any scientific hypothesis a heterogeneous body of information that cannot be stated explicitly in full. A judgment of prior probability gives at least a rough assessment of the way that knowledge applies to a given hypothesis. (Salmon 1990, 331)

What about that assessment then? To warrant his apparent confidence with respect to the accuracy of the assessment, Salmon adds

> It is, to be sure, a subjective judgment, but it also reflects objective experience. It is important that any such prior probability assessment be made in the light of scientific experience, to the exclusion of idiosyncrasies, prejudices, ideologies, and emotions. (Salmon 1990, 331)

Certainly! However, Salmon mounted his argument in this paper starting with a concession that it is not just a body of information, or the content of objective experience, that plays a role here, but also judgements about the other virtues. So we still have the same question before us: in that case, is it still appropriate to read the probability involved as probability of truth, rather than degree of belief? Is there an additional basis for taking the degrees of belief, when they were formed with that additional bias in favour of those virtues, to be tracking the truth? Salmon does not address those questions.

It appears therefore that in developing his argument Salmon's response to the challenge was to slide from one question to another one. Putatively, he is addressing the role of admittedly external theoretical virtues in theory choice, but when it comes to the point, he offers a justification that leaves them out of account.

3) Ivanova makes an important new move in the debate about theoretical virtues by bringing in Tania Lombrozo's studies of actual human reasoning. This is an important, quite new, field of study in cognitive psychology. I would like to suggest that the work by Lombrozo's colleague and co-author Alison Gopnik be taken into account as well, specifically to illuminate the role of our universal impetus toward explanatory conclusions. Gopnik's article "Explanation as Orgasm" (1998) does not trade on any easy analogy, such as to an "aha!" experience. The analogy she displays points to a structural similarity between two natural processes in human affairs:

> My hypothesis will be that explanation is to cognition as orgasm is to reproduction. It is the phenomenological mark of the fulfillment of an evolutionarily determined drive. From our phenomenological point of view, it may seem to us that we construct and use theories in order to achieve explanation or have sex in order to achieve orgasm. From an evolutionary point of view, however, the relation is reversed, we experience orgasms and explanations to ensure that we make babies and theories. (Gopnik 1998, 102)

Scientific theorizing could in principle progress without explanations. As far as that is concerned, though, we might never have progressed beyond mere data-fitting. Fortunately there is something driving us toward creativity, speculation about what is behind the scenes, explanation beyond what we can verify; our metaphysical instincts have a crucial historical and psychological role. The philosopher's delicate task is to distinguish what makes something more likely to be true from what makes it more attractive to us.

Reply to Michael Friedman, "Realism and Instrumentalism: The Best of Both"

Michael Friedman addresses the recurrent question of whether there is a via media for scientific realism and anti-realism that will preserve the best of both. In this attempt he tries to clarify Howard Stein's contention to have achieved this, something that indeed needs clarifying. Does Friedman succeed? His conclusion, illuminating about how far Kant still reaches, does not include a precise statement of the Steinian position with which, as he says, he takes his stand.

But once we realize that the crux of Friedman's paper is the appeal to George Smith's concept of theory-mediated measurement, to resolve the issue between Kant and Newton, it becomes clear just where the realism/anti-realism—or as I prefer to say, realism/empiricism—controversy is being joined.

I accept completely that the main insight into Newton's methodology gained in recent decades is the realization that it involves centrally and crucially *theory-mediated* measurement. The questions at issue are therefore: how can the scientific realist ratify this methodology, and how can the empiricist accommodate its value?

Friedman takes the example that Kant introduces in a critical comment about Newton's "deductions". Kant's comment becomes understandable with Friedman's explanation of it, and so does the quite complex example of planets' acceleration induced by their satellites. That is valuable, but I think that we can get a clearer, or at least more reader-friendly, idea of the implications for the epistemology of science by looking at a simpler example. That is the experimental set-up designed by the Reverent George Atwood, presented in his book *A Treatise on the Rectilinear Motion and Rotation of Bodies, with a Description of Original Experiments Relative to the Subject* (1784).

This set-up is now usually called the Atwood machine, and we can parse his description on pages 299–300 this way:

> The Machine consists of two boxes, which can be filled with matter, connected by a string over a pulley. The ideal case, modelled most easily, has an inextensible massless string, and the pulley is massless, with zero friction retarding the motion, which occurs in a vacuum.
>
> *Result:* In the case of certain matter placed in the boxes, the machine is in neutral equilibrium regardless of the position of the boxes; in all other cases, both boxes experience uniform acceleration, with the same magnitude but opposite in direction.

Reverent Atwood 's purpose was adversarial, as he shows in this reference to the continuing influence of Cartesian physics on the continent:

> Many experiments, however, have been produced, as tending to disprove the Newtonian measure of the quantities of motion communicated to bodies, and to establish another measure instead of it, viz. the square of the velocity and quantity of matter; and it immediately belongs to the present subject, to examine whether the conclusions which have been drawn from these experiments arise from any inconsistency between the Newtonian measures of force and matter of fact, or whether these conclusions are not ill founded, and should be attributed to a partial examination of the subject: but some considerations concerning the principles of retarded motions should premised. (Atwood 1784, 30)

To put this in context it should be appreciated that Cartesian physics did not die with Descartes, and that Newton's theory too had to struggle for survival, for almost a century. The Cartesian critique of Newtonian physics was that by introducing *mass* and *force*, which are not definable in terms of spatial and temporal extension (they are, we say now, dynamic rather than kinematic quantities) Newton had brought back the medievals' occult qualities. For, according to the Cartesians, only quantities of extension are measurable.

The Newtonian's response was, in effect, that admittedly what is measured *directly* in any set-up is lengths and durations, but that they could show nevertheless how to measure mass and force. But the calculation of the mass ratio between the two filled boxes relies on principles in Newton's theory.

So how can the Cartesian see this as evidential support for Newton's theory, rather than as a fallacy, begging the question, since some of the premises come from Newton's theory?[10]

Briefly, the deduction goes like this: the accelerations of the two boxes, with masses M and m, are equal but opposite in direction. Newton's theory implies that magnitude of the acceleration is proportional to the ratio (M − m)/(M + m), which determines the ratio M/m. Hence measuring the acceleration of the boxes tells us, upon calculation, their mass ratio. But the calculation comes from Newton's own theory.

Well, then, the Atwood experiment does not *confirm* Newton's theory. But it is still precisely what the Newtonian has to show: that this experiment, given that its results in all its instances are coherent with each other and with Newton's theory of mass and force, *supports* Newton's theory. There is no question that it does so, and no question that this is precisely the sort of support a theory which involves quantities that are not directly measurable, requires.

So for the epistemology of science, this is the conclusion, and sets us a task: to distinguish confirmation (*making more likely to be true*) and experimental support, via theory-mediated measurement, and to make the distinction clear.

Scientific realists and empiricists have, in a way, the same task here, but may view it differently. If realists wish to argue that such experimental support does nevertheless, despite the air of petitio principii, confirm the theory, they will have to provide the requisite argument.

10 The example of theory-mediated measurement of mass was at the heart of the nineteenth-century reconsideration of the foundations of mechanics. On the assumption that the apparatus is located in a uniform gravitational field, without needing to know the strength of that field, the rationale of this response was thoroughly re-investigated in the nineteenth and early twentieth century by Mach, Duhem, and Poincaré (Mach 1960, Ch. II, Section V-1–3; Poincaré 1952 [1905], 97–105). Some conclusions can indeed be drawn without presupposing Newton's theory: as Mach points out (*The Science of Mechanics* Ch. II, Section I.16) Atwood's machine shows, and allows to measure directly, the constant acceleration postulated in Galileo's law of falling bodies. But the measurement of Newton's dynamic parameters on a body is an operation that counts as such a measurement only relative to Newtonian theory. To say that the operation measures mass, for example, is to presuppose the applicability of Newton's second and/or third law.

As to the empiricist, as I have argued elsewhere (van Fraassen 2012 and forthcoming), theory-mediated measurement is at the heart of the *empirical grounding* of a theory, in ways originally spelled out by Hermann Weyl. From the empiricist (and I would say, from a practical) point of view, the truth of theories does not matter, their empirical grounding matters very much.

Reply to Matthias Egg, "Stances and Doctrines in Scientific Metaphysics"

In an interesting turn of events James Ladyman characterized "scientific metaphysics" as a stance rather than a thesis, theory, or doctrine: the scientistic stance.

Ladyman argued correctly that the empiricist stance and the materialist stance are compatible, and could be combined—but that point needs to be qualified. It is correct in the minimal form in which those stances are presented in *The Empirical Stance* (van Fraassen 2002). Materialism was there diagnosed as consisting of a cluster of attitudes that include strong deference to the current content of science in matters of opinion about what there is, as well as an inclination, or perhaps commitment, to accept approximative, suitably qualified, completeness claims for the current sciences. And it involves a readiness to adjust opinion so as to follow suit as the sciences change. All of this characterizes Ladyman and Ross's weak metaphysics, who add an acute critique of traditional "strong" metaphysics similar to the empiricist critique. Finally, they share what was presented as the positive side of empiricism, the admiration of scientific forms and practices of inquiry, which make the sciences a paradigm of rational inquiry into empirical matters of fact.

That in Ladyman and Ross' attitudes the materialist predominates is clear enough, if we keep in mind that for the materialist the sciences teach us what to believe, but for the "mere" empiricist science is more nearly what teaches us how to give up our beliefs.

But as Matthias Egg clearly shows, Ladyman and Ross did not remain with the minimal form in which that tension might be negligible. As they develop their position the beliefs involved go well beyond what the sciences could teach. Hence the aptness of referring to their position as a scientific metaphysics or naturalistic metaphysics.

We should clearly not carp about such terms as "metaphysics", which are circumscribed very differently in different philosophical traditions. In my own critique of metaphysics I was targeting what Ladyman and Ross call strong metaphysics, and recognize full well that this traded on a very narrow sense of the word.

The question which Egg addresses is whether their scientistic stance is tenable in the form they give it, and more specifically whether it is subject to internal strains, and eventually whether it avoids incoherence. This is a delicate enterprise, for evaluation of a philosophical stance can always be vulnerable to the charge that it is evaluation from within a different stance, trading on different values, and therefore biased.

This is a salient issue in Anjan Chakravartty's discussion, and Egg is entirely right, it seems to me, to point out that Chakravartty undermines his own argument with his broad, liberal conception of rationality. The criterion of rationality for philosophers taking a stance is that it does not involve self-sabotage by their own lights. This is an internal, stance-transcendent criterion. Note well that meeting or failing to meet this criterion is entirely independent of whether it is possible or impossible for someone taking a different stance to judge whether or not the criterion is met. Perhaps it is impossible for us to say with certainty that some other stance is incoherent, in that or any other sense, given the possibility of presuppositions that we have but have not uncovered. But even if that is so, it would be a fallacy to infer that there is no sense to the notion of incoherence. The fallacy would be similar to the inference from unverifiability to a lack of cognitive significance or of truth-value.

Egg offers an incisive critique of the claim that the scientistic stance is a coherent—let alone "perfect"—combination of empiricism and materialism, with reference to the measurement problem in the philosophy of quantum mechanics. I would like to say more about that, in a moment. But first, I would like to add that Ladyman and Ross offered several mutually contradictory characterizations of their own enterprise. So in their book, on page 2, they write about basing a metaphysics on science:

> Since this knowledge can be incorporated into unified pictures, we also can have some justified metaphysics. Based as it is on incomplete science, this metaphysics probably is not true. However, if it is at least motivated by our most careful science at time t, then it is the best metaphysics we can have at t. (Ladyman and Ross 2007, 2)

In what sense of "best"? If it is probably not true, then it is not best in the sense of being most likely to be true. It might be the most explanatory, but—not being true—of things that aren't so. Is the claim then that it is the most likely, among metaphysics in progress, to evolve into a true theory of everything? Against this they warn us on page 9 that

> careful work by various philosophers of science has shown us that this task is not straightforward because science, usually and perhaps always, underdetermines the metaphysical answers we are seeking. (Ladyman and Ross 2007, 9)

If that is what we have to look forward to (which seems likely to me!) the words "justified" and "best" do not sound apt. And if it is a warning they mean to heed, it comes just a bit late since on page 7, answering Jonathan Lowe's critique of naturalism, they can argue that on their side "the metaphysical assumptions in question are vindicated by the success of science" (Ladyman and Ross 2007, 7).

There is much valuable insight to find in their book, and I do appreciate the greater understanding of scientific theories that we obtain whenever careful imaginative interpretations are presented. But it belies their own meta-philosophical statements of purpose when they align that with the traditional metaphysical purport to be a pursuit of truth.

There is, as Egg comments, a very intriguing side to the scientistic stance which attributes to science a life-orienting role. Negatively, linking philosophy so closely to the sciences will direct people away from confabulated alternative world-pictures and from self-destructive policies. Positively, metaphysics is a source of inspiration for further empirical research as well as further theorizing, leading researchers in new directions for fruitful work. A metaphysics based on science provides us with a genuine world-picture, an account of the world and our place in it. Empiricism cannot claim to do that, only metaphysics can claim to have successfully usurped the role that religion had before the advent of this thoroughly secular, Enlightenment stage in Western civilization.

Well, that is all so. But if I may quote the same passages again, that even this naturalistic metaphysics is based "as it is on incomplete science, this metaphysics probably is not true", and what we can expect from scientific progress is that "science, usually and perhaps always, underdetermines the metaphysical answers".

So this additional value of the scientistic stance can perhaps be appreciated as a value only within the scientistic stance. From another vantage point it gives the appearance of a self-referential paradox, counselling belief in what you do not believe.

Egg ends by addressing the question of coherence for the scientistic stance by raising it for how Ladyman and Ross respond to the quantum-mechanical measurement problem. That is not a problem for an empiricist view of quantum mechanics. Whether the theory is empirically adequate is independent of whether there are interpretations of the theory in which measurement outcomes are derivable from the postulated dynamics, with or without hidden variables. The representational role of the theory is important, and certainly not to be denied. But a correct understanding of representation, and of contextual criteria of accuracy for representation, will disarm the philosophical aporiae.

Can the same be said for the scientistic stance Ladyman presents? Egg has argued convincingly that "[d]issolving the measurement problem is not an option for the realist" (Egg 2019, 62). The scientific realism which Ladyman and Ross aim to maintain within their stance must maintain the representational role of the quantum-mechanical formalism, in a way that goes beyond what the empiricist grants. To give a simple analogy, think of Martin Schongauer's engraving of the temptation of St. Anthony, in which the saint is surrounded by flying devils. That is certainly a representation of the saint's situation, with meaningful elements. But whether it is an accurate representation of what a contemporary of the saint would have seen if looking at him—perhaps it is!—is independent of whether there existed anything corresponding to the flying devils in the painting. The painting tells a coherent story which explains the agonized look on the saint's face. But, for someone taking a realist position with respect to that story, if I may paraphrase Egg, dissolving the devils' reality problem is not an option. As above, we see here as well that the desire, that their position be not just a stance but also truly a naturalistic metaphysics, leads Ladyman and Ross right into the same problems as beset the traditional "strong" metaphysics.

Reply to Nancy Cartwright, "Evidence, Relevance and Warrant: In Defence of Voluntarism"

There is much resistance to voluntarism in epistemology, and it seems that the case for voluntarism needs to be argued again and again. In the seventeenth century both Descartes and Pascal realized that any claims to knowledge, or to a basis for belief, are threatened by a regress, and offered us the main opposing positions. Descartes concluded that the regress stops at apodictic, self-evident truths while Pascal concluded that knowledge has no foundation, that we are always in our epistemic and doxastic life engaged in a wager.

In the subsequent history, and perhaps especially over the last century or so, we have seen a remarkable disconnect between epistemology and what happens in scientific practice, but also—where Nancy Cartwright plays a central role—sustained efforts to close the gap. We are of one mind here, we differ only in that her arguments have been much more practice-directed.

Foundationalism in epistemology, or what is currently called evidentialism, is the rival tradition that needs to be countered at every point. As Cartwright mentions, there was the *myth of the given*, which both of us (being students of Wilfrid Sellars) take to have been completely discredited. Yet there remains, it seems to me, everywhere in general epistemology, a reflexive turn to the subjective to look for warrant or certainty. This is most visible in the examples that still illustrate arguments there—"I seem to see a bird", "I feel a pain", even if not Russell's "red here now". These examples are strikingly disconcerting, given that what counts as evidence in the sciences, or in a court of law for that matter, is first and foremost required to be something public, and publicly accessible, and must at least in principle be testable or evaluable by everyone.

This bias toward the subjective goes well beyond what is unconsciously revealed in the choice of examples. Timothy Williamson wrote that my evidence is all and only what I know. But my claims to knowledge have no status as evidence, and the content of my knowledge does so only if you or anyone can ascertain that content as well. Roger White, when his earlier arguments against voluntarism in epistemology (what he calls "epistemic permissiveness") were challenged, announced that by "my evidence" he meant "all that I have to go on". Leaving aside that the phrase "to go on" is just as endorsing as "known", and that the objective status of what a person takes herself to have to go on is not in the least guaranteed to be transparent to that person, this line shows the same bias toward the personal and subjective that is absent from any serious presentation or evaluation of evidence in legal or scientific practice. Contrast these naïve sounding abstractions, that introduce something so flimsy to serve as a foundation of sorts, with Cartwright's argument theory of evidence, constructed explicitly in the conviction that there is no foundation to be found.

Through her alignment with Otto Neurath's philosophy Cartwright sides with Pascal, insisting that our epistemic and doxastic life has no foundation. The regress stops only by being stopped at a chosen point, by *plumping*, that is, by a free decision. But she backs this up with a detailed analysis of what happens in such an act, when done responsibly and with scientific objectivity. As she points out, the very notion of evidence sends us, upon reflection, in several different directions. There is no such thing as evidence *überhaupt*, "evidence" is an implicitly polyadic term. There is only evidence *for*, and this being *evidence for* is itself a relation that is relative—relative to what? ... *to an argument*, in her spelling out. It does not stop there, for the argument will have premises that themselves need warrant, as much as the warrant that the argument itself provides for the evidence claim, and the warrant will involve values as well as further facts for substantiation.

Cartwright rightly emphasizes that central in any such argument will be assertions of relevance, of what bears on what. Judgements of relevance, however much they appeal to facts, are value judgements. That is clear in science-guided practical decision making: in the cost-benefit analysis, was it taken as relevant whether the costs would be are borne by those to whom benefits would accrue, and does it differentiate between risks taken-on voluntarily and risks imposed?

But controversies in the foundations of physics offer equally salient examples. Those begin in the seventeenth century when Galileo declares as solely relevant the common sensibles, what came to be called the primary quantities, and Descartes' subsequent restricting those further to just quantifiable qualities of extension. Though unsustainably strict, those voluntarily, deliberately self-imposed limits played a significant role in the early development of modern science. To choose a much more recent example, in assessing the theoretical import of violations of Bell's Inequality in the recordable data, is there a relevant possibility of a pre-established cosmic harmony between the settings of the apparatus and the timing of the experiment? Although this is typically, and usually just implicitly, dismissed, it has been recurrently pointed out that without eliminating this possibility the usual argument from those violations is not logically valid. Taking the stance that these quantum phenomena establish a radical, previously unconceived indeterminism in nature can therefore only be seen as a choice, revealing a value judgement, even a worldview.

Cartwright's reasoning, with its detailed analysis of evidential argumentation in practice, goes well beyond the more general, and hence more superficial, reasons for voluntarism that I have offered. I would just add something more in support, because I am not as dismissive of probabilist models of belief and opinion as Cartwright (van Fraassen 2022). Her reflection on the inevitable regress, which can in the end only be engaged by plumping, is entirely applicable to probabilistic thinking. That supports her contention, and broadens its scope. My telling example—I call it *telling* because it is immediately convincing in the telling—is the (possibly apocryphal) quotation from famous rock climber Lynn Hill: "What I need to assess on any occasion is, first, the probability of a fall, second, the damage of a fall, and third, my own ability to assess the first two." There is no escaping assessment of assessment, nor the unavoidable call to decision as to when to call an end to it.

Finally, the issue of volition in our doxastic/epistemic life is not a bare, bloodless matter of academic dispute, it implies an inseparability of epistemology and ethics. I want to emphasize, as Cartwright does in her conclusion, the ethical consequences: "Nevertheless, we are responsible for our actions." The responsibility for what we believe, how we frame or manage our opinions, or what we take as basis for our decisions, rests solidly on our shoulders.

References

Atwood, G. 1784. *A Treatise on the Rectilinear Motion and Rotation of Bodies, with a Description of Original Experiments Relative to the Subject.* Cambridge. University of Cambridge. Online at https://wellcomecollection.org/works/kctnz9s3/items?canvas=9 (last accessed: 24.11.2023).

Cartwright, N. D. 2007. Why Be Hanged Even for a Lamb? In: Monton (2007), 32–45.

Dellsén, F. 2017. Reconstructed Empiricism. *Acta Analytica* 32: 95–113.

Egg, M. 2019. Dissolving the Measurement Problem Is Not an Option for the Realist. *Studies in History and Philosophy of Modern Physics* 66: 62–68.

Gopnik, A. 1998. Explanation as Orgasm. *Minds and Machines* 8: 101–118.
Kant, I. 1977. *Werke in zwölf Bänden.* Ed. W. Weischedel. Vol. 9. Frankfurt am Main: Suhrkamp.
Kant, I. 1998 [1781/1787]. *Critique of Pure Reason.* Trans. and eds. P. Guyer and A. W. Wood. Cambridge: Cambridge University Press.
Kant, I. 2004 [1786]. *Metaphysical Foundations of Natural Science.* Ed. M. Friedman. Cambridge: Cambridge University Press.
Lambert, K. 2002. Henri Lauener: A Very Personal Rememberance. *Dialectica* 56(4): 295–296.
Mach, E. 1960. *The Science of Mechanics.* Trans. T. J. McCormack. LaSalle, IL: Open Court.
Massimi, M. 2017. What Is This Thing Called 'Scientific Knowledge'? Kant on Imaginary Standpoints and the Regulative Role of Reason. *Kant Yearbook* 9: 63–83.
McAllister, J. 1996. *Beauty and Revolution in Science.* Ithaca, NY: Cornell University Press.
Monton, B. 2007. *Images of Empiricism: Essays on Science and Stances, with a Reply from Bas van Fraassen.* Oxford: Oxford University Press.
Percival, P. 2007. An Empiricist Critique of Constructive Empiricism: The Aim of Science. In: Monton (2007), 83–116.
Poincaré, H. 1952 [1905]. *Science and Hypothesis.* New York: Dover.
Salmon, W. 1990. The Appraisal of Theories: Kuhn Meets Bayes. *PSA: Proceedings of the Biennial Meeting of the Philosophy of Science Association* 1990, Vol. 2: 325–332.
van Fraassen, B. C. 1980. *The Scientific Image.* Oxford: Oxford University Press.
van Fraassen, B. C. 1993. From Vicious Circle to Infinite Regress, and Back Again. In: *PSA: Proceedings of the Biennial Meeting of the Philosophy of Science Association* 1992, Vol. 2: 6–29.
van Fraassen, B. C. 2002. *The Empirical Stance.* New Haven: Yale University Press.
van Fraassen, B. C. 2005. The Day of the Dolphins: Puzzling Over Epistemic Partnership. In: *Mistakes of Reason: Essays in Honour of John Woods*, eds. A. Irvine and K. Peacock, 111–133. Toronto: University of Toronto Press.
van Fraassen, B. C. 2012. Modeling and Measurement: The Criterion of Empirical Grounding. *Philosophy of Science* 79: 773–784.
van Fraassen, B. C. 2014. One or Two Gentle Remarks about Hans Halvorson's Critique of the Semantic View. *Philosophy of Science* 81: 276–283.
van Fraassen, B. C. 2019. How Can We Understand Transcendence of the Ego? *Nova et Vetera* 17(2): 77–93. http://dx.doi.org/10.13140/RG.2.2.33488.97284 (last accessed: 10. 9. 2022).
van Fraassen, B. C. 2022. Reflection and Conditionalization: Comments on Michael Rescorla. *Nous.* https://doi.org/10.1111/nous.12416 (last accessed: 24. 11. 2023).
van Fraassen, B. C. (forthcoming). Theories, Modeling, and Empirical Support. In: *The Modeling of Energy Transition: Cultures—Visions—Narratives*, eds. R. M. Erdbeer, V. Hagenmeyer and K. Stierstorfer. Basingstoke: Palgrave Macmillan.

Contributors

Claus Beisbart is Professor for Philosophy of Science at the University of Bern. There, he is not only affiliated with the Institute of Philosophy, but also a member of the Oeschger Centre for Climate Change Research and the Center for Artificial Intelligence in Medicine and Co-Director of the Multidisciplinary Center for Infectious Diseases. He holds a PhD in physics and another in philosophy, both from the Ludwig Maximilian University Munich. From 2022 to 2023, he has been President of the Swiss Philosophical Society. He has been Co-Editor of the *Journal for General Philosophy of Science* from 2015 to 2020. His current research focuses on computer-based methods in science, in particular computer simulation and machine learning, philosophy of cosmology and philosophical methodology. He is the co-editor of the Handbook *Computer Simulation Validation* (Cham 2019, with N. J. Saam).

Nancy Cartwright is a philosopher of science, specializing first in the philosophy of physics and later, on the social and economic sciences, and recently on improving policy deliberation. She is Professor of Philosophy at Durham University and Distinguished Professor at the University of California San Diego, having worked previously at Maryland, Stanford and the London School of Economics. Cartwright is a former MacArthur fellow, a fellow of the British Academy and the (UK) Academy of Social Sciences, a member of the American Academy of Arts and Sciences, the American Philosophical Society, the Academia Europeae and Leopoldina. She has won the Hempel and (with Elliott Sober) Lebowitz Prizes. Her latest book is *A Philosopher Looks at Science* (Cambridge 2022).

Finnur Dellsén is Professor of Philosophy at the University of Iceland, in Reykjavík, Iceland, and Professor II at Inland Norway University of Applied Sciences, in Lillehammer, Norway. He holds a PhD in philosophy from the University of North Carolina at Chapel Hill. He is the recipient of the Lauener Prize for Up-and-Coming Philosophers in 2018, and the Nils Klim Prize for young Nordic scholars in the humanities, social science, law, and theology in 2019. Most of his current work focuses on (a) how Inference to the Best Explanation can be made precise and sophisticated in a probabilistic epistemological framework, (b) the nature of scientific understanding and its role in science, particularly with regard to scientific progress, (c) disagreement in science and its implications for the public perception of science.

Matthias Egg is lecturer and researcher in philosophy at the University of Bern. He holds a MSc in theoretical physics from the University of Zurich and a PhD in philosophy from the University of Lausanne. He was a visiting scholar at the Universities of Notre Dame and Leeds, and has served as substitute professor in philosophy at the Universities of Lausanne and Wuppertal. His postdoctoral research, collected under the title *Quantum Reality and the Prospects of Scientific Metaphysics* resulted in a habilitation at the University of Bern in 2022. His current research is situated at the intersection of philosophy of science and philosophical anthropology, with some inspiration from the history of philosophy and theology. He is the author of *Scientific Realism in Particle Physics: A Causal Approach* (Boston/Berlin 2014).

Michael Frauchiger, Dr. phil., has been lecturing in the fields of philosophy and methodology of science and the humanities with the University of Applied Sciences of Zurich, the Open University and the University of Bern, and additionally has been an Advanced Research Fellow of the Swiss National Science Foundation. Since its establishment in 2003, he has been the Managing member of the board of trustees of the Lauener Foundation for Analytical Philosophy. His research cuts across epistemology, philosophy of language and logics, metaphysics, philosophy of religion, philosophy of action, ethics and political philosophy. He has edited or co-edited all volumes published in De Gruyter's series *The Lauener Library of Analytical Philosophy.*

Steven French was until recently Professor of Philosophy of Science at the University of Leeds. He gained his PhD from the University of London and held positions in Brazil and the USA before moving to Leeds. He is a former President of the British Society for the Philosophy of Science and Co-Editor-in-Chief of the *British Journal for the Philosophy of Science* and is currently a member of the Académie Internationale de Philosophie des Sciences and also of the Academia Europaea. His main research areas are in the philosophy of physics and the philosophy of science, more generally, and his most recent book is *There Are No Such Things as Theories* (Oxford 2020). In recent years he has been examining the Husserlian stance towards physics, culminating in his final work, *A Phenomenological Approach to Quantum Mechanics: Cutting the Chain of Correlations* (Oxford 2024).

Michael Friedman is Patrick Suppes Professor of the Philosophy of Science at Stanford and director of the Stanford Suppes Center for Logic, Methodology, and Philosophy of Science. He is a member of the American Academy of Arts and Sciences. He was the recipient of the Franklin J. Matchette Prize of the APA (1985), the Lakatos Award in Philosophy of Science (1987), a Humboldt Research Award (2008), the Fernando Gil International Prize in the Philosophy of Science (2015) and presented the De Gruyter Kant Lecture at the Central Division of the APA (2021). His current research fields are the history of the philosophy of science, Kant and his successors, history of analytic philosophy, and more. Among his many books the most recent is *Kant's Construction of Nature: A Reading of the Metaphysical Foundations of Natural Science* (Cambridge 2013). He is co-editor (with R. Creath) of *The Cambridge Companion to Carnap* (Cambridge 2007), and editor of William Demopoulos' *On Theories* (Cambridge, MA, 2022).

Milena Ivanova is a Bye-Fellow at Fitzwilliam College at the University of Cambridge, UK. She holds a PhD in philosophy from the University of Bristol. She has held positions at the Munich Centre for Mathematical Philosophy at the Ludwig Maximilian University of Munich and the University of Sydney. She was also a visiting fellow at the University of Queensland. Her work investigates how aesthetic values shape scientific practice, with her current focus being on the aesthetic dimensions of scientific experimentation. She is the author of *Duhem and Holism* (Cambridge 2021) and co-editor of *The Aesthetics of Science: Beauty, Imagination and Understanding* (London, 2020, with S. French). Currently, she is co-editing a book on *The Aesthetics of Scientific Experiments* and working on a monograph titled *Beautiful Science.*

Michela Massimi is Professor of Philosophy of Science at the University of Edinburgh. She gained a PhD in philosophy from the London School of Economics. She was Junior Research Fellow at Girton College, Cambridge, and taught in the Department of Science and Technology Studies at the University College London before moving to Edinburgh in 2012. She has held visiting positions in Pittsburgh and Dartmouth College. Massimi is a Fellow of the Royal Society of Edinburgh, of the Royal Astronomical Society and Member of the Academia Europaea. She has written extensively in history and philosophy of science. More recently, she completed a monograph entitled *Perspectival Realism* (New York 2022).

Bas C. van Fraassen is Professor Emeritus at Princeton University and Distinguished Professor at San Francisco State University, and has held previous positions at Yale University and the University of Toronto. He has honorary degrees from the University of Lethbridge, University of Notre Dame, and the University of Leuven, and is a Fellow of the American Academy of Arts and Sciences, Foreign Member of the Royal Netherlands Academy of Arts and Sciences, and Corresponding Member of the British Academy. Honours include the Lakatos Award, the International Giulio Preti Prize, and the Inaugural Hempel Award. His Locke Lectures at the University of Oxford became the 2008 book *Scientific Representation: Paradoxes of Perspective* (Oxford).

Index

https://doi.org/10.1515/9783111019802-015